THE CORE CHAIN

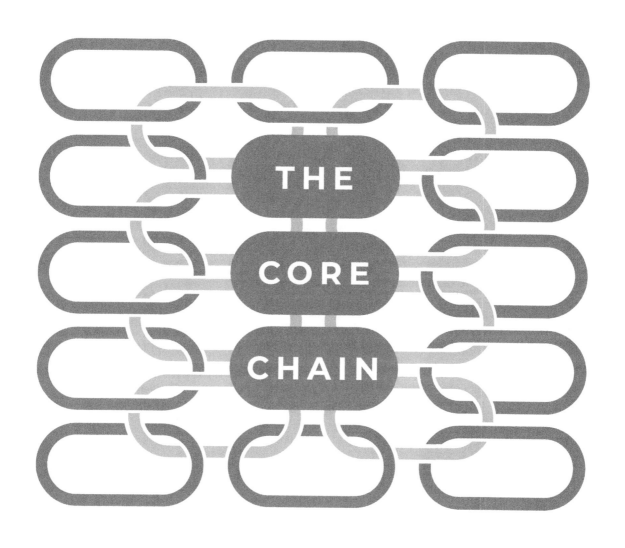

A Survival Guide to the Perilous World of Medical Device Development

JUSTIN PAUR

Copyright © 2023 Justin Paur

All rights reserved. No part of this book may be reproduced in any manner without express written consent of the author except in cases of brief quotations or excerpts for use in reviews and critiques.

Hardcover: ISBN 979-8-9874138-0-7
Paperback: ISBN 979-8-9874138-1-4

Editing by: Richard Dionne, Barry Lyons
Art by: Justin Paur, Anna Hryshchenko, Made Wirawan
Cover art by: Paul Palmer-Edwards

ACKNOWLEDGMENTS

Thank you to the whiteboard watchers, the individuals who sat in classrooms and conference rooms and let me work through my half-baked ideas. Each one of you provided patience, insight, and encouragement, which helped evolve an idea to a concept, then a concept to a methodology. I count you all among my friends and advisors, and forever will. Randy Nelson, Kristin Mortenson, Matt Stoll, Georgiann Keyport, Thom Gunderson, Jon Schell, Lee Swanson, Carole Norris, Steve Parente, Carla Pavone, and Jim Bracke.

Thank you to the fixers, the beta-readers, the red-liners, the book editors. Only two of you were paid, and every one of you were vital to the progression of this book. It is hard to read an unpolished book, especially one about medical device development. You were never as mean as I asked you to be, which speaks to how wonderful all of you are. Steve Gray, Randy Nelson, Richard Dionne, and Barry Lyons.

Thank you to Robert Langer, Susan Alpert, and Manny Villafaña. Despite your status, busy schedules, and many other commitments, you took the time to sit with me and share your wisdom. I am truly grateful.

Thank you to my parents, who spilled more red ink on this book than anyone else. You are wonderful parents, and as with most kids, I can never thank you enough or pay for everything I broke.

Thank you to my wife Aundi. You were a whiteboard watcher, fixer, advisor, and much, much more. Your support was steadfast from the beginning, even though my estimations of the commitment were, let's say, *optimistic*. This would not be remotely possible without you. You are the love of my life, and I am the luckiest guy I know.

Thank you to my four-year-old, Brooks, who would have sat by my side during every hour of writing this book if I'd have let him. Nobody else can evoke such love, terror, and laughter in such quick succession. You are the best thing. At bedtime you often tell me you love me infinity plus eight. I love you the exact same amount.

No thanks for my dog, Winston. You were incredibly distracting during this entire experience.

CONTENTS

Mountaineering 1

PART I: DEVELOPING A MEDICAL DEVICE USING TRADITIONAL TECHNIQUES 7

1. Welcome Aboard 9
2. Traditional Product Development—Stage 0, Ideation 11
3. Traditional Product Development—Stage 1, Concepting 15
4. Traditional Product Development—Stage 2, Product Development 19
5. Traditional Product Development—Stage 3, Design V&V 23
6. Traditional Product Development—Stage 4, Pivotal Trial 27
7. Traditional Product Development—Stage 5, Launch/Post Market 31
8. Addressing Concerns 33
9. Going Lean 35
10. Closing Shop 37
11. Failure Rates 39
12. The Postmortem 45
13. Part I: Recap 47

PART II: THE CORE CHAIN FRAMEWORK 49

14. The Point of What We're Doing 51
15. Making It 61
16. Tying It All Together 67
17. What We're Building 79
18. Part II: Recap 81

PART III: THE CORE VALUES 83

19. Core Value #1—Build the Right Chain 85
20. Core Value #2—Build from Left to Right 93
21. Core Value #3—Prevent Translation Errors 97
22. Core Value #4—Build Looking Left, Check Looking Right 101
23. Core Value #5—Delineate, Delineate, Delineate 105
24. Core Value #6—The Smaller the Links, the Tougher the Chain 109

25. Core Value #7—Rough It in and Refine 113
26. Core Value #8—Check and Protect 117
27. Part III: Recap 123

PART IV: THE CORE CHAIN METHOD OF PRODUCT DEVELOPMENT 125
28. Welcome Aboard 127
29. Introduction to The Core Chain Method 129
30. The Core Chain Method—Stage 0 141
31. The Core Chain Method—Stage 1 151
32. The Core Chain Method—Stage 2 159
33. The Core Chain Method—Stage 3 167
34. The Core Chain Method—Stage 4 173
35. Part IV: Recap 177

Conclusion 181
Frequently Asked Questions 185
Glossary of Acronyms and Terms 189

MOUNTAINEERING

Mountain climbing sounds fun, doesn't it? The crisp, clean air filling your lungs. The natural beauty surrounding you. The overpriced but comfortable gear you're draped in. I've always pictured it as a leisurely stroll in the wilderness, if only slightly uphill. It's more about a dogged perseverance than any real technique. Just bring enough water and peanut butter sandwiches to get you through it. And a carabiner, I think.

I'm originally from North Dakota, probably one of the flattest locations in the world. We have a saying about North Dakota: "It's the only state where you can see your dog run away for three days." My formative visions of mountains mainly stemmed from action movies and the occasional inclines and declines of central Minnesota. So, when I moved to the Appalachian Mountains of West Virginia to finish my undergraduate studies, I gained a completely different respect for mountains.

The Appalachian Mountains are ever-present, looming, and begging to be explored. And, from a safe distance, climbing even the largest of them seems achievable. Looking at them day in and day out from the comfort of my classroom, I began to gain confidence in my imagined climbing abilities. I had played high school sports after all, never mind that it was for a school that was too small to require any sort of try-out. Once my confidence had reached a critical mass, I committed myself to venturing out. I embarked one Saturday to a small local mountain, finding a well-groomed path for what I assumed would be an effortless climb. I strode onto the trail with a skip in my step and a new walking stick in hand. Five hours later I barely reemerged, battered and bruised, sans walking stick. I shuffled to my car, where I cursed my hubris and licked my wounds. I couldn't walk without leg cramps for the next three days. I had been thoroughly humbled by the smallest of mountains.

One of my classmates was a West Virginia local. I told him about my mountain climbing experience, and how it was quite a bit more difficult than it looked. "You weren't mountain climbing, mate," he told me. "You were mountain hiking. Big difference." He chuckled a bit before continuing. "Mountain hiking is a stroll in the woods. You park, you walk up, then you walk down, and you have a beer in the parking lot. Mountain climbing, on the other hand, is serious.

It involves risk, proper training, equipment, and a clear and well-executed route. Climbing a mountain can take weeks, and you need to understand exactly what you're getting into before setting out." Noticing my deflated expression, he continued, clearly enjoying himself. "That's not even the half of it," he said. "There's another level to this, and that's *mountaineering*. That's for the difficult summits, and it's quite advanced. That's a lifestyle, and it's a dangerous one. You, my friend, are a mountain hiker." As our school mascot was modeled after the rugged mountaineer, I took some offense to that conclusion. But with more research, and a cursory glance at my squishy-soft physique, I decided he was exactly right.

• • •

Any product development endeavor is an uphill battle. It takes planning, ability, grit, and, of course, money. The size of the hill, or the mountain for that matter, can be dramatically different based on the industry you're working in. Some endeavors require that leisurely but invigorating stroll I was originally envisioning; some require the mountain hiking that almost did me in. Medical device development, however, is not mountain strolling or mountain hiking. It is not mountain climbing. It is *mountaineering*.

If you're new to medical device development, welcome, and please don't be scared off by this analogy. The experience of bringing a medical device to market is an excellent one, and it's well worth the effort. But I want to make sure you know exactly what's ahead of you before you set out. Medical device development is not like conquering the rolling foothills in Arizona or the small mountains in West Virginia. We are pursuing the big peaks. The ones where the summit extends far past the clouds. Maybe not Mount Everest or K2; that honor goes to our friends in Pharma. But nonetheless, these peaks ahead of us are quite serious, and are not to be taken lightly.

As with mountaineering, medical device development is about following the best route to the summit while employing the right tools, training, and talent for the terrain. The most successful mountaineers possess an unnaturally deep reserve of determination, while being magicians at making their resources last. The same can be said for the most successful medtech developers. The ones that make it to the top can often seem supernatural to others. They are not supernatural, however. They know, as we do, that just a few steps off the most efficient path can spell disaster.

True mountaineers also watch out for one another, considering themselves part of a community. Nobody cheers when someone fails in their efforts. While competition and business pressures are present, we in medical device technology are lucky enough to be working in the rarified air of lifesaving and life-improving technologies. When our industry is successful, and more companies reach their summit, the world directly benefits. Lives are saved, quality of life is improved, and our ability to help each other goes up a notch. That's why it's important that we are not only good at this as individuals, but as a community.

This book is about making us better at developing medical device technology. In an industry where mistakes are costly, there is a surprising lack of clarity about exactly what we're building. Without clarity, we can wander off the path, fall into the same traps as our industry colleagues, and leave critical technologies on the table.

Here is where the *core chain framework* comes in. Traditional medical device product development is filled with inconsistencies, self-induced burden, and, by its fundamental structure, waste. Using the core chain framework, along with the core values (principles to follow), we can rebuild the product development methodology from the ground up.

The core chain method is a new approach to medical device development that is structurally different than previous techniques, providing a clearer picture of what we're building. This new structure is better aligned with the patient, inherently less wasteful, and speeds up the product development process. Speed and quality often determine success, after all, and this method will get more of us to market with the same resources in play.

As much as I'm chomping at the bit to introduce you to the core chain, I have learned that it isn't best to just "jump in." To truly understand the state of development in the industry, what the issues are, and how we can improve, we must all have the same foundational knowledge. To that end, this book will start by introducing the traditional medical device development method, often called "the waterfall method." This is the current standard of medical device development across the industry, and anyone serious about this industry should understand it. The "waterfall" in the waterfall method refers to a diagram created by the FDA[1]:

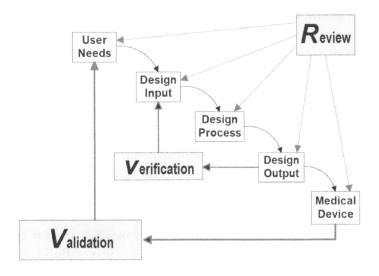

FDA Design Control Waterfall Process

---

1   From the "Design Control Guidance for Medical Device Manufacturers," FDA, 1997.

THE CORE CHAIN

Most product development methods in the industry are based on this graphic, organized in a stage-gate methodology, with one or two of these waterfall elements built into each stage. While the number of stages sometimes changes, the basic structure is the same. We will broadly call this type of development the "waterfall method."

The waterfall method performs two functions: (1) it acts as a mental framework for tying individual activities together, and (2) provides an organizational system for these various activities. Eventually, we are going to tweak both the framework and organizational system, but we must first internalize this waterfall method. To that end, you and I will go through the exercise of bringing a new medical device to market, showing the waterfall method, and briefly introducing individual development *activities* along the way. Please pay attention to the word "briefly" here. You will notice that there is a paragraph or two on each of these activities, when people can (and have) written entire books about just one activity. As we proceed through the exercise of bringing a medical device to market, we'll also hear feedback about this process from some of the industry's foremost figures: Manny Villafaña, Robert Langer, and Susan Alpert. We'll then examine the common failure modes of medical device companies, and what that means for the future of medical device development. This section, developing a medical device using the waterfall method and describing the individual development activities, is **Part I** of this book.

Once we feel comfortable with the general activities needed to bring a medical device to market, along with the waterfall method itself, we will step back for a minute and reconsider our perspective. We will evaluate the clinical problem we are trying to solve, along with the product we are trying to develop, now from the patient's, caregiver's, and regulator's perspectives. We will use these perspectives to help us build the *core chain*, which will reframe our product development activities. The core chain framework will seem quite simple, even intuitive, when laid out. But make no mistake, it has far-reaching consequences, not only to the organizational system for the product development activities, but to the individual activities themselves. This section, building and describing the core chain, is **Part II** of this book.

With the core chain understood, we will start to explore the consequences of this framework, and how we can use it to avoid the traps into which many product development companies fall. We will synthesize these lessons into our eight *core values*. Each core value will provide a principle to follow when developing any medical device. These core values leverage the core chain framework and help simplify and realign the product development process with the patient and caregivers. While these core values are simple in theory, they have proven to be crucial in practice. Describing the core values and their implications is **Part III** of this book.

While the core chain provides the fundamental framework, and the core values provide the principles to follow, we can and will go one large step further. Building on the core chain and the core values, we will introduce an entirely new method of medical device development, named the *core chain method*. Here, we will reconsider the exercise of bringing a medical device to market from Part I, but this time with the core chain, the core values, and the core chain method in hand. We will see how using this methodology fundamentally changes the approach to product development, and how, compared to the waterfall method, we are able to speed up development while reducing our resource needs. This, then, introducing the core chain method, is **Part IV** of this book.

Each one of these elements is intended to help us on our journey. The core chain provides a clear picture of where we are going; the core chain method lays out the most efficient path from beginning to end; and the core values help keep you from wandering into the wilderness. When used together, we can make our efforts cleaner, faster, and better aligned to the patient. This keeps us tight to our path while optimizing our resources, giving us the best shot at our summits.

That's it. Sounds like fun, eh? I hope you enjoy the book, but more importantly, I hope you can internalize the concepts presented within. As you read on, please remember that I'm not an author (until now, I guess), so if there are sections that are clunky or jokes that swing and miss, I appreciate your grace.

As we dive headfirst into the weeds, there are a couple more items I'd like you to keep in mind to get the most out of this book—especially for you industry veterans.

First, please don't get caught up in nomenclature. In this industry there are a lot of different names for things. Where I say, "product specifications," you may say engineering specifications, design specifications, product design specifications, or any other variation on the theme. While I initially assumed that regardless of nomenclature, everyone would be on the same page, I found I was completely wrong. People in our industry tend to get caught up in nomenclature. So, rather than trying to account for every variation of each name, I picked what I found to be either the most common or the clearest, then did my best to describe exactly what that name represented. Please do your best to focus on that description, then reapply your name of choice. Also, there is a glossary in the back of the book for your reference if you get lost in the alphabet soup that is our industry's acronyms.

Second, while many strategies and methodologies are proposed in this book, you can and should make them your own. The medical device industry is broad and deep, and attempting to make a map for one situation tends to neglect others. Luckily, the broader themes I propose will still apply, so it is up to you and those around you to challenge and improve these ideas, creating the best methodology for your situation. Play with these ideas, mold them as you like, and give your development method a snappy acronym, because who doesn't like those?

The rest we'll figure out as we go, and if you have thoughts or questions that I don't address by the end, please jot them down and send me a passive-aggressive email. That's how we Minnesotans prefer feedback.

PART I

DEVELOPING A MEDICAL DEVICE USING TRADITIONAL TECHNIQUES

Part I	Part II	Part III	Part IV
Developing a Medical Device Using Traditional Techniques	The Core Chain Framework	The Core Values	The Core Chain Method of Product Development

In Part I, you will join my start-up company, and we will go through the process of developing a medical device using the traditional waterfall method. Along the way we will hear about the difficulties of this approach and the common pitfalls from industry leaders.

1
WELCOME ABOARD

We're so thrilled you decided to join our new start-up! Welcome to your first day at our little company. I know you're as excited as we are to bring this new medical technology to market. We've been anxiously awaiting your start date, and we feel we can officially start our product development activities now that you're on board. I know today is your first day and there is a lot to do, but as you are part of the leadership group, I think you and I should spend the morning together.

Now, during your interviews we gave you an introduction to the idea we're hoping to commercialize. I'm guessing it's a big reason you decided to join us. Of course, we're excited about it too, but we have a long way to go before we can get that idea into the hands of the clinicians. Generally, as part of the leadership team, you'd be expected to understand how we intend to bring this exciting idea to market, as well as help to execute that development process every step of the way. But as you are our new head of research and development, you'll be expected to not only understand the process, but to lead it. To that end, I want to make sure you're as prepared as you can be. So, why don't you get comfy, maybe grab a double-latte, and let's start diving into the first stage of our product development journey.

2

TRADITIONAL PRODUCT DEVELOPMENT—STAGE 0, IDEATION

Now, as always, time and resources will be tight, so we'll need to have the most effective product development process we can muster. Luckily, you and I are old pros at this, so we'll use the *traditional waterfall methodology* for our product development process. We both know this methodology well, but as we need to stay closely aligned, let's walk through each stage of the process as we get into it, starting with the stage we're in now. I understand that you already know much of what I'm telling you, and in some cases, far more, so I appreciate you humoring me. We are going to need to explain this process many times to people both inside and outside of this company, so let's make sure the path in front of us is as clear as we can make it.

Let's start from the beginning. So, officially, welcome to **Stage 0: Ideation**.

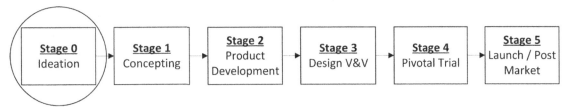

This is the stage where ideas are born. As ideas sit in people's heads or in lab books, this stage often lasts years. That was true enough for us, but finally our idea has matured enough for us to officially start the development process.

Here's an overview of what we'll be doing in this stage:

THE CORE CHAIN

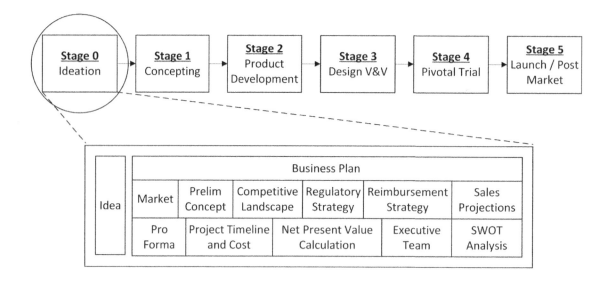

In this stage we're spending most of our time putting together one thing, the business plan. While this is considered a document, it has many separate parts that need to be created independently. We will put them together into a clear narrative that shows the entire picture of what we're proposing to do, and the landscape for the product. We would create this even if we didn't need external funding, but as we do need outside money, this business plan must instill confidence in people who are not as familiar with the details as we are. Here are some of the major elements of this document:

Market—We will identify and describe the disease state and resulting population we intend to target. We will reduce that number by identifying the portion of the market that we can possibly address. We will then describe the competitive landscape that is already addressing that market, and the pros and cons of those technologies.

Preliminary Concept—We'll need to describe our technology and how it will work. Any pedigree or background information can be added here too (if it was developed in a secret high-technology lab, for instance). A rendering of the concept would be a bonus, but at the very least, we'll need some sort of visual so our potential investors can get a sense of what we're thinking and planning.

Competitive Landscape—While our idea is unique, we're probably not the only ones attempting to help this market. We'll need to provide the relevant patent landscape and associated freedom to operate in the design space we're using. The more thorough we are here, the better. Our ability to patent and protect our device is a big part of what we're selling to investors. We'll include United States and global filings.

Regulatory Strategy—As we will be launching a medical device, we will need to involve the appropriate regulatory bodies. For the United States, it will be the Center for Devices and Radiological Health (CDRH) at the FDA. We will need to assess the likely device classification (Class I, Class II, or Class III), and whether we will be submitting a Premarket Approval (PMA) or 510(k) (the submittal for a product that is substantially equivalent to an already approved device). As part of this, we'll also identify our clinical trial strategy. As our device will likely have a clinical trial, it will end up being a major part of our overall costs and timeline. It'll make sense to bring in a regulatory consultant to confirm our assessments.

Reimbursement Strategy—Some companies miss this, but it is quickly becoming a major component of the medical device product development process. Even if the FDA, patients, and doctors want this product, it will probably die a quick death if the insurance companies are not willing to pay for it. Will this product fall under established reimbursement codes, or do new ones need to be created? Does this add or remove costs from the healthcare system? If it adds cost, what is the tangible quality-of-life benefit? What will the major payors need to see to approve this product? We need to address *all* these questions in this section.

Sales Projections—Using market, competitive landscape, and reimbursement strategy sections, we'll build up a projection of sales once we launch and for the first years thereafter. We need to think this through and show that our numbers have actual meat behind them. Every time someone justifies their sales projections through "the market is xx billion and we only need to capture one percent of that," I want to go find a wall to bang my head against. We will describe our sales channels, required sales force, distribution networks, and so on. We'll analyze poor/good/great scenarios to show the possible range of sales we're expecting.

Pro Forma—The economics of the company, once we reach commercial distribution, are what investors are ultimately betting on. We will take the sales projections, likely cost of goods sold, gross margin, selling and administrative expenses, and create our resulting net profit for each quarter and year. We will do a sensitivity analysis to show what happens in the poor/good/great sales scenarios from an overall company economics perspective.

Project Timeline and Cost—Until we go commercial, money will only flow one way: out. We have to show how much cash we need and when we need it. We will create a high-level project plan with time and cost of each major step. This will include resources required (both from a financial and personnel standpoint), what the project milestones will be, and the estimated time until launch. The quality and level-headedness of this project plan will help tell investors that we know what we're doing, and what we'll need to be successful.

Net Present Value Calculation—At this point we will have our estimated costs and our estimated income over the life of the product. This will allow us to create a net present value (NPV) calculation. We'll show the NPV based on our plans, but we will also scenario out the poor/

good/great cases for market adoption as well as NPV if the project is delayed or overspends. We will also go one step farther and show a risk-adjusted net present value (rNPV), which will adjust the NPV down based on a risk factor.

Executive Team—This section matters a lot to investors, as they want to know that we are not just playing at this, and that we know what the heck we're doing. Think of this as a line-up card for a baseball team. We need to show how the various roles required for this type of endeavor are met, and the strength of each individual in their position. I just bought a new suit, so I get to be the CEO, and of course you are the head of research and development. We also have a regulatory expert and a finance and marketing director on the team. The bigger and more seasoned the team, the more comfortable the investors will be, to a point. We also don't want 10 expensive people on the payroll with little for them to do.

SWOT Analysis—One of the last things they might want to see is a strengths, weaknesses, opportunities, and threats (SWOT) analysis. This is often a good place to discuss the various competitive and regulatory risks as well as the potential for a strategic-level company to swoop in and buy us out. Projected timelines are often also a little, let's say, "rosy" for start-ups, so we will put in an analysis of what will happen if we're delayed in hitting our milestones.

We may sprinkle in some other sections in the business plan, but that will be the bulk of it. We'll also compress it down into a slide deck that we can go through in about 25 minutes. Then we'll start setting up pitch meetings with potential investors.

That's it for Stage 0. Let's get ready to woo some investors.

3

TRADITIONAL PRODUCT DEVELOPMENT— STAGE 1, CONCEPTING

Well, we wooed the investors. Good work. They liked our business plan well enough, and decided to write us some checks. Our seed round was officially funded, so we can keep the lights on for now. Our investors had some hard questions, but in the end, they trusted us and our analyses. The money came with some strings attached, so from now on we'll have an official board that will act as the gatekeeper between the remaining stages. Hopefully, they'll help keep the financial wolves at bay so we can focus on our product. To that end, back to work. Welcome to **Stage 1: Concepting**.

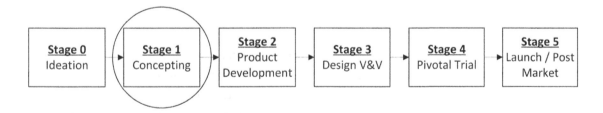

This is probably the most delightful stage in the process. Future versions of ourselves will look back on this period as fun and fancy-free. As you might have guessed from the name, the main activity in this stage is making and refining product concepts. We'll get into this, but know that concepting is not all we'll be doing here. This stage moves pretty fast, so let's review what we need to accomplish:

THE CORE CHAIN

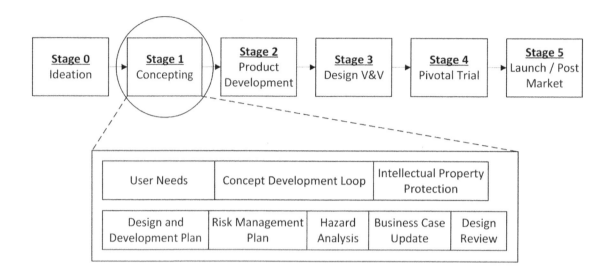

User Needs—The user needs (aka market requirements) allow us to describe our target patient's and caregiver's needs, converting them into actual requirements for the product. We must craft these carefully, as once they are complete, we will lock them down and control revisions. These user needs, and their revision history, will become part of the **design history file (DHF)**.

Concept Development Loop—Get your popsicle sticks and pipe cleaners ready. During these early stages our concepts will feel like kindergarten craft time, but don't worry, we'll get fancy later. We will only create concepts that test what we need them to test. No reason to machine it out of titanium when modeling clay will do the job. We will first use this to visualize and cement our ideas, then we'll start testing them against our requirements.

Intellectual Property Protection—We will start to protect any concepts we deem feasible. We're going to keep our patent lawyer busy for a while as we try to capture most ways to make a product like ours. Our patents will be provisional to start, and we can convert them to full patent applications when we are farther down the development path. If our idea is good (and it is), there will be others trying to take a piece of it, so we need to block the space as best we can.

Design and Development Plan—We took a stab at the project plan for the business case in Stage 0, but now we need to translate it to an official design and development plan. We will describe the process we're following to bring our product to market, along with the deliverables needed for each stage along the way.

Risk Management Plan—Risk management is a critical part of the medical device development process, and it's important that we define how we will address it. We will describe how we will evaluate and categorize risk, and what documents we will deliver as part of the risk management process.

Hazard and Harms Analysis—As the first of our many risk management deliverables, we will perform a hazards and harms analysis. We will document the risks this product poses to the patient, as well as the foreseeable ways the product could be misused by the caregiver.

Business Case Update—We will reexamine the business case at the end of every stage, leveraging our new experiences. We will revisit and expand on each of the major sections in our original business case. It will take some serious digging to revise some of these items, and throughout this stage we will need to bring in experts and external services to refine our estimates.

At the end of this stage, we'll want at least one solid concept that we feel we can run with, along with justification for that confidence: our testing, intellectual property (IP) landscape, and technical risks predicted with this concept. We will also perform our first **design review,** which we will do at the end of each stage moving forward. The design review will cover the user needs, hazard analysis, design and development plan, and risk management plan. As part of the design review, we will need to sign on the dotted line that the device is ready to advance to the next stage in our development.

Now that we have a board, we will also have a gate review before they allow us into the next stage. In the gate review, they will challenge what we've done up to this point. If they're smart (and they are), they will hold the line, not letting us pass until we've accomplished what we need to accomplish, both in the eyes of the regulators and the eyes of the business. They know we have just put our toes into the water so far; in the next stage, we will be cannonballing into the deep end. Our plan must be rock solid, our funding secured, and our team prepared before the board will let us through.

4

TRADITIONAL PRODUCT DEVELOPMENT– STAGE 2, PRODUCT DEVELOPMENT

Thank you for attending the Stage 1 gate meeting with us. They can be a bit rough. You probably noticed me sweating bullets as they picked apart every minor weakness in our plan and made us answer for our assumptions. I guess that's their job. Anyway, they liked the concept and the landscape enough, so with a couple of tweaks to the business plan they are passing us through to the next stage. Take a breath, we made it. Welcome to **Stage 2: Product Development.**

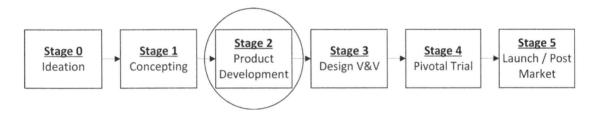

Stage 2 is when we truly get to flex our engineering muscles and get into the nitty-gritty of the design. This is also a stage where our team will start growing. In addition to our little army of engineers, we have started to add quality and regulatory specialists. I'll get into why they're here in a bit. Soon I'll be calling an all-hands meeting to lay out our next activities, but let's make sure you and I are aligned first. Here's what we'll be up to in this stage:

THE CORE CHAIN

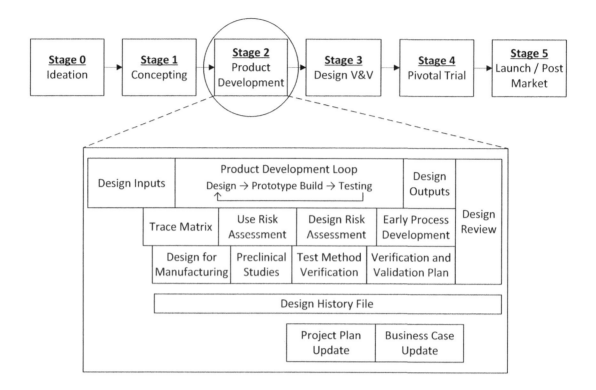

Design Inputs Development—To guide us in the design process we will need to create a set of design inputs (aka system requirements). In the last stage we created our user needs. We will now try to convert those into more tangible inputs to our design. These design inputs will become the standard by which we measure all our design iterations. Once we have these, we can kick into our product development iteration loop.

Product Development Iteration Loop—This is the meat and potatoes of this stage and will likely encompass the vast majority of our dollars and time for a while. The primary loop will be design → build → test → redesign (back to the beginning). Here's the detail:

Design—We will start with the concepts we developed in the last stage, then we'll formalize the design we intend to test. We will start the process with very general, rudimentary designs and dial it in on each iteration.

Prototype Build—We will need to physically build what we design, and there are a few ways to do this. As with the design, we'll start simple and get more sophisticated as we get closer and closer to the final product.

Testing—As the medical device development guru Randy Nelson puts it: at this stage you *test to failure*. In the next stage we'll *test to success*. The tests should challenge the product against the design inputs and/or the user needs. We will document the results of each test on each design iteration to show how our different ideas have performed and progressed. While this will be less formal than after we do a "design freeze," we'll still capture it all and document it in the design history file.

Design Outputs—At the end of this stage we will need to provide a set of "frozen" engineering specifications. These are our design outputs, and they take a few forms, but they are often a mix of engineering prints and specification documents. These documents will dictate the materials of construction, the physical dimensions (with allowable tolerances), how components fit together, and so on. Strict revision control starts here as well, often illustrated by the change of revision numbering schemes (for example, we are no longer on revision six, we are on revision A).

Trace Matrix—With the user needs and design inputs created, we will develop a trace matrix that shows which design inputs trace to which user needs, then ultimately, which design outputs trace to which design inputs. Eventually, we will add our product verification and validation activities to this trace matrix to show how different elements trace together.

Use Risk Assessment (uFMECA)—The use risk assessment will assume that the product was designed and manufactured correctly and will consider all the ways that the product could be *used* incorrectly by the user and/or patient. What would happen if they put it in backwards?

Design Risk Assessment (dFMECA)—As the design begins to come into focus, we will need to conduct a full design risk assessment. This risk assessment will help inform our design process, establishing whether we need to reconsider aspects of our product to mitigate a serious risk to the patient. In making this document, we can look forward to a few long days of being stuck in a room filled with spreadsheets, and various forms of sugar and caffeine. This document becomes the basis for a lot of the decision-making that will occur downstream of this event. It is considered a living document (although revision-controlled), so figure on updating as we learn.

Early Process Development—If there are higher-risk processes or new technologies that we need to employ, we'll give them special attention. For each of these processes, we will identify the major risks and create a plan to test and develop the processes. The process development will continue in the next stages, but in this stage, we will need to know if it's not feasible to build the product as designed.

Design for Manufacturing—Through this process, we will have to keep in mind that the set of skills needed to make a handful of devices is not the same as the set of skills needed to fill a warehouse with devices. We will bring in general manufacturing and automation experts, as well as experts in each individual process. We will also get deeper pricing information from them (both unit and capital pricing) to help dial in our cost of goods and project cost projections.

Preclinical Studies—As part of our development testing in this stage, we may start our preclinical studies, which will require official protocols. These studies may include benchtop and animal studies, and will help guide our design efforts as well as help establish the safety profile of the device.

Test Method Verification—If you can't properly test the product, you can't verify or validate it. We need to ensure that the tests we are using are actually providing us useful information about the product's attributes and potential failure modes. We will verify that we have selected the appropriate tests and that they provide meaningful data.

Design Verification and Validation Plan—In the next stage, we will vigorously test most aspects of the design. We will need to be able to justify that each test we perform is suitable for what we are trying to learn. By leveraging the testing we performed in the product development iteration loop, we'll start to build up the final verification and validation (V&V) test plan.

Design Review—When we believe we have accomplished what we need to accomplish in this stage, we will have a formal design review. Key people will sit in a room again, reviewing the design outputs and risk assessments, ensuring that they meet requirements. Everyone will again sign their name on the dotted line, indicating that the product is ready to move on to the next stage.

Design History File—This file (aka the DHF) is intended to tell the story about how the product progressed from idea to finished product. The revision and design controls allow visibility into how the product was imagined, how it was made, how it performed on tests, and how the product evolved as a result. Many of the activities occurring in this stage will end up in this file, and if it is launched in the United States, this file will be submitted to the FDA.

Update of the Project Plan and Business Case—As we approach the end of the product development stage, it will be time to update both the project plan and business case. Our investors/gatekeepers will want to know how we are performing against our plans, and whether our previous assumptions/analyses still hold true. They won't hold true, of course; it's more a matter of how *off* we were, and in which direction.

Depending on how things go, we may be in this stage for six months, or we may be in this stage for three years, but we will not progress into the next stage until we are ready. The gatekeepers may bring some independent technical analysis to judge the worthiness of the design as well as our development work, so we'll need to keep things sharp. Once we're ready, we'll be able to put an actual product in their hands that we think will help a lot of people. Until then, we'll stay diligent as we keep moving the product along.

5

TRADITIONAL PRODUCT DEVELOPMENT— STAGE 3, DESIGN V&V

Alright, we now have a product that we feel is going to work. The board agrees, although they had some reservations. They always do. It's all quite subjective up to this point anyway; the real tests are coming after all. You're going to be busy in this stage testing our product six ways from Sunday, and I'll have to do another funding round to build up the war chest for this and the horribly expensive stage coming up. Settle in because we're going to be here for a while. Welcome to **Stage 3: Design Verification and Validation**.

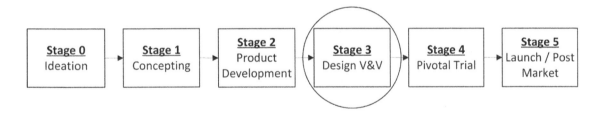

The design verification and validation (Design V&V) stage, above all else, is about testing. We need to ensure that this product is fully vetted prior to human use. In order to achieve this, we are going to perform a large suite of tests. Some of the tests will be standards that are used for most medical devices, and some tests will be very specific to our product and application. There are some other areas we're going to be building up in this stage as well. Here's the lay of the land:

THE CORE CHAIN

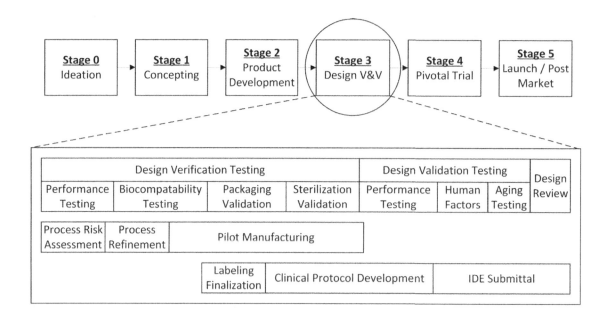

Design Verification Testing—These tests match your design outputs with your design inputs. We may have performed some of these tests during the previous stage, but we now have a set of locked design outputs/engineering specs, so we can officially test them against the design inputs. We'll need protocols and reports for these tests. Here are some of the tests that we're going to be performing as part of the design verification activities:

Performance Testing—We need to ensure that the various components of the device function as intended, in a simulated clinical environment. These tests aim to verify that the various elements of the device fulfill the design input requirements.

Biocompatibility Testing—We are going to make sure the materials we chose don't cause an adverse reaction by the body. You might want to rethink those asbestos racing stripes you added.

Packaging Validation—The packaging you designed must protect the product throughout our distribution channels until it reaches the patient. We don't want our product poking through the side of the tray and breaking the sterile barrier of the device before the user receives it. We need to test the packaging through its worst-case distribution conditions, including the winters of North Dakota, summers of Florida, and the potholes of Indiana.

Sterilization Validation—Since our product will be labeled "sterile," we need to validate that it is indeed that. We need to ensure that we are manufacturing it in a facility that controls the bioburden on the product (that is, our cleanroom), and that the sterilization cycle kills the bugs everywhere on the device.

Design Validation Testing—The other suite of tests, the design validation activities, will test whether we built a final device that actually fulfills the needs of the user. Again, we will follow the protocol-execution-report format, and our activities should follow the design V&V plan. We need to use production-equivalent product, so we will need to do some pilot manufacturing to make the product we need to use for this testing. As you know, we may get into good laboratory practice (GLP) animal studies and a pivotal clinical trial (next stage) as part of our validation testing. Following are three of the design validation tests we will likely need to perform:

Performance Testing—We will validate that our product performs as intended (per the user needs) under actual or simulated or real-use conditions. This may include animal studies, or even human feasibility clinical trials.

Human Factors (Usability)—Here we need to see how the actual users of our device interact with our device without our assistance. With only the instructions for use and other labeling to guide them, we'll see if they get confused, use it improperly, or use it in unexpected ways.

Aging Testing—Also known as *shelf life* testing, we need to take some of the product we made and determine if it loses its safety and efficacy profile over time. We will submit the product to accelerated and real-time aging conditions, then we will re-execute some of the product-performance tests to ensure that the product performs throughout our shelf life.

Process Risk Assessment (pFMECA)—This is the third of the "big three" risk analyses in the product development process. We will dig into each process and define the ways in which they can go wrong and decide what to do about it. Risk assessments are always fun activities where you argue semantics for three days straight.

Process Refinement—Here we will start to explore the various processes further and understand what the critical variables likely are. Once we know our "vital few" variables, we'll then roll into our process characterization activities, where we'll test the variables and create process windows where we know the process will result in acceptable product.

Pilot Manufacturing—There are different levels of this activity, all dependent on what we need for testing. For validation testing, we're going to need product that is "substantially" equivalent to the final product made on the final equipment. We need to use the same processes we will use in commercial manufacturing (although likely at a different scale).

Labeling Finalization—We will be locking down the labeling in this stage in preparation for the clinical trial. The FDA has a say in it, so we can't call it done until they give the thumbs-up. Every word, symbol, and font size will be scrutinized, so we don't want to mess this up. Incorrect labeling is a major cause of recalls, so let's get it right.

Clinical Protocol Development—The clinical trial is an extension of the design validation activities, so it makes sense to prepare the protocol at this stage. We will show the FDA the

THE CORE CHAIN

design of the clinical trial we intend to perform. We could spend months talking about different clinical study designs, so we'll bring in experts to help us find the best path forward based on our device. In general, we will define the population and disease state we're targeting, what the "end points" of the study will be (how we show safety and efficacy), and what statistical evidence we intend to provide at the end of the study. This is an extremely delicate matter, so we must craft carefully. Many effective products have not made it past the FDA because of poorly designed clinical trials.

Design Review—After the design validation activities, we will perform another design review, ensuring that the final product meets the user needs. This review is one last kick of the tires before we start putting these devices into people for our final design validation activity, the clinical trial. Once again, we will document our approval, list any outstanding items, and add the design review meeting minutes to the design history file (DHF).

Investigational Device Exemption (IDE) Submittal—Once we have the verification and validation data, along with the clinical protocol, we will submit the IDE to the CDRH (Center for Devices and Radiological Health) at the FDA (government agencies love acronyms). We will march forward with preparations, but we will not start using our product on humans until the FDA gives us the thumbs-up.

The verification and validation stage is a stressful one for the team, but we need to make sure that our product is properly designed prior to the pivotal trial. Many of the sins we committed during the design process will come back to bite us in this stage, so be prepared. Sound like fun? I think so. Big smiles.

6

TRADITIONAL PRODUCT DEVELOPMENT– STAGE 4, PIVOTAL TRIAL

We have submitted our IDE to the FDA. High five! We cannot start the clinical trial until we get the go-ahead from them, so we will be waiting for a bit. Let's get ourselves mentally prepared for this stage. Once we use our device in actual patients, we are in a completely different ball game. Welcome to **Stage 4: Pivotal Trial**.

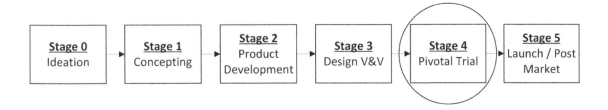

Now we will be waiting on a response from the FDA not once, but twice in this stage, so it's hard to predict how long this stage will last. Stage 4 also contains our pivotal trial, so we'll have to be generally flexible with our timing. The regulatory side should (and will) set the cadence of the activities in this stage. The other departments just need to make sure they don't get in the way. A passing of the baton will occur as well, as the R&D engineers take a backseat to the operations engineers. Here are the activities we can expect:

THE CORE CHAIN

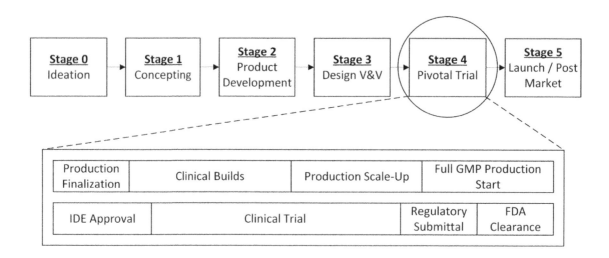

Process Validation/Design Transfer—Our next build is going to be for our pivotal trial, so we need to answer any outstanding questions and validate our production prior to the clinical build. This holds true both at our facility and at our vendors' facilities. We will also complete the design transfer activities, snapshot the production, and end with a first article run to ensure we're making good product.

Clinical Build—The clinical build is a big deal for operations, and they should know the gravity of the build they are performing. Everything prior to this point had theoretical risk; now these manufacturing builds will carry actual risk to patients. If we have any questions or hesitations with our manufacturing, we need them cleared up before we start building. We will build the first product prior to IDE approval and hold on to it until we gain the green light from the FDA. We'll then continue to build through the first part of the clinical trial, resupplying the trial as it continues.

Production Scale-Up—Our core manufacturing processes will be locked at this point, but we don't have the capacity we'll need for launch. So, we will scale up our manufacturing. We will need to buy, install, and validate larger, faster (and more expensive) equipment. Getting the equipment alone will take six months, so we need to get this started prior to getting approval from the FDA for commercial production.

Full GMP Production Start—Once we feel FDA clearance is likely, or after we receive it, we will start full GMP (Good Manufacturing Practices) production. This is our commercial production, and if we have regulatory approval, these units are saleable.

IDE Approval/Clinical Trial—As soon as we get IDE approval (and right before, ideally), the clinical trial preparation and execution will dominate our company's activities. The clinical trial is a big effort, and it is where most of our funding will be spent. We will be outsourcing the clinical trial management to a clinical research organization (CRO). We will have some experts in-house overseeing the activities, but we will use our CRO to execute the plan. There are many regulations and requirements around clinical trial management, and it'll take a lot of effort to prepare, educate, and audit clinical sites, gather and store HIPAA data, perform interim analysis, manage adverse event reporting, and so on.

Regulatory Submittal—Once we have compiled all the data from the design history file, clinical evaluation report, device master record, and a whole bunch of other activities, we will submit that to the FDA—in our case, in the form of a 510(k) submittal. This submittal used to involve pallet-loads of paper, including multiple copies in case they wanted to read it twice. This (surprisingly recently) is no longer the case. The FDA can now handle electronic copies exclusively. We're also lucky we don't have to submit a PMA (Premarket Approval), which we'd have to do if we had a higher-risk device or one without a strong predicate (a previously approved comparable device). Those can be brutal, with only a few handfuls of products going through that process each year.

While the review time for a 510(k) is 90 days, you can count on receiving questions from the FDA, which extends that time frame. It changes every year, but, as a rule of thumb, the average time for a 510(k) approval is about three to five months. For a PMA, the review time is 180 days, but average approval time is actually around seven to nine months.

FDA Clearance—Eventually, the FDA will give us a response. Wanna make a bet that the response from the FDA is a set of questions? Four-to-one odds . . . Ah, I know you're not taking that bet. Once they send us the questions, we will get them the answers, and their ticking clock will resume. We will stay in this stage until we get that FDA clearance for commercial launch, and we will continue to spend money with no revenue to offset it. If this takes too long, the company will need to start hibernating while we wait.

Our kindly investors know that there may be a little back-and-forth with the FDA. Our models suggested this could happen, so they have patience, for now. While we have a minute to breathe, let's clean up some of our outstanding issues and start to prepare for other markets. Get comfy, we may be here a while.

7

TRADITIONAL PRODUCT DEVELOPMENT— STAGE 5, LAUNCH/POST MARKET

WELL, WE HAVEN'T yet received a "go" for commercial launch from the FDA. They have some questions that we are working diligently to answer, but it will take a while to get things in order. I am as confident as ever that we, and our technology, will prevail. With that in mind, let's look ahead to our next stage in preparation for our approval.

Next is hopefully the longest and most beneficial stage, what we've been working hard to reach for the last few years of our life. So, let's take a peek at **Stage 5: Launch**.

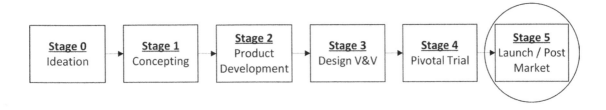

Stage 5 is very different from the others. Our product development team will be celebrating, but also likely busy with the next product. Our operations team will be deeply involved for the duration and will be cursing any mistakes made by the product development team. We are no longer developing the product; we are repeatedly building what we developed, over and over, hopefully for years and years. The project plans we created in Stage 0 and Stage 1 all end here. There is no fixed end to this stage until we decide it's time to retire the product.

THE CORE CHAIN

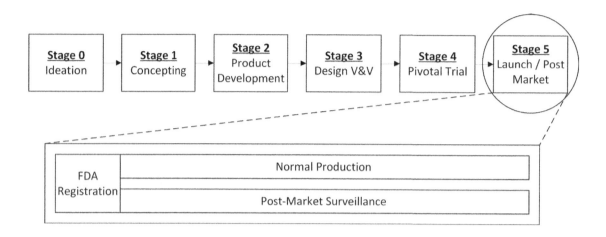

FDA Registration—First things first: when we get approval from the FDA, we'll need to be registered as the manufacturer. This is a relatively simple process since we have the documentation associated with the 510(k). But know that as soon as we register, it is very possible that our friends at the FDA will stop over for a friendly visit (aka audit).

Normal Production—Normal production is just what it sounds like: normal production. We will build the product according to the specifications we created, fulfilling the requests from sales. Where this gets tricky is change and issue management. We'll have a combination of planned changes (changes to the process that we'd like to make for various reasons), and unplanned changes (something isn't working as expected). We will handle these through our corrective and preventative action (CAPA) system, nonconformance (NC) system, and through change control. We will need to bring in new equipment for capacity or replacement reasons, keep our equipment calibrated, our people trained, docs up to date, and our facility in good working order. I won't get too far into details here, but know that while we defined everything up front, production lives and breathes, and we'll need to work hard to keep everything in order.

Post-Market Surveillance—Adverse event tracking, complaint tracking, and returned product handling systems will all need to be in place by the time we launch. This is very important and will be a natural point of scrutiny for the FDA. We made some claims about the safety and efficacy of our device, and it's important that we ensure that the product is behaving as we expected. We are required to notify the FDA of adverse events and to investigate what happened and why. Many aspects of the medical device development process are debatable. This one is not. We will do a good job at this, or the FDA will have no qualms about yanking the rug out from under us.

Alright, the investors are breathing down our necks to find out what is happening with the FDA. We have a meeting coming up so we will get the lay of the land soon.

8

ADDRESSING CONCERNS

Well, the FDA is saying that we haven't fully proven the safety and efficacy of our device. Can you believe it? We believe our data is good enough, but they want us to perform some additional clinical trials. We have spent considerable time and effort pushing back, but they are sticking by their original assessment. So for now, we're stuck in Stage 4.

I, along with the clinical team and our advisors, will put a plan together to address their concerns with a small confirmatory trial. We will submit the proposal and get feedback from the FDA. Hold tight because this may take a minute. In the meantime, please continue your efforts on the production preparation, but you may want to slow your activities a bit. We are, of course, on a hiring freeze, but your team will stay intact for now.

9

GOING LEAN

This is taking longer than expected. The regulators do not believe that the small clinical trial we are proposing will adequately address what they say are major concerns. Our investors are not willing to pay for another large trial. So, we're getting creative. We are bringing in some specialist statisticians to see if there are other ways to get the statistical proof we need, or if there is a subset of the overall population against which we can make our product claims. We are also talking to some strategic-level companies to see if they are willing to step in and bring this across the finish line. Hold strong but go lean. Please stop all manufacturing operations and layoff your nonessential personnel. Stage 4 is starting to feel like purgatory. We're going to get through this.

10

CLOSING SHOP

Please pull up a chair and close the door. I'm guessing you know what I'm about to say. Well, this is a sad day. The investors are forcing us to close shop and liquidate. They are saying that even with another round of funding, they don't think we're going to make it to market. I can hardly blame them. The FDA doesn't seem excited about our product, and even if they did allow it through, our clinical data isn't terribly compelling. That is the most difficult part for me. I know the need is there, and I know that the product can really help, but the data seems to disagree.

This is tough, but stiff upper lip and all that. I will be announcing that we are laying off all employees as of today and selling off our assets to the highest bidder. I hope that means that our technology still has a chance to make it to the market, but it doesn't look promising. While I am disappointed in our fate, the real shame is that the doctors and patients will likely not get access to a technology that can really help.

Well, thanks for coming along for the ride with us. Sorry for the bumps, and the crash landing, I guess. At least we gave it the old college try. We're headed out to a pub to commiserate before we go our separate ways. Please join us, first round's on me.

11

FAILURE RATES

Thank you for going through that exercise with me. I decided to walk through the waterfall product development process with what is unfortunately the most likely outcome. Medical technology companies sometimes fail, and that's okay. Some companies failing is not the problem. In reality, *most* medtech start-ups fail, with some estimates putting the failure rate as high as three-quarters.[2] That's less okay. Failure is so common, in fact, that the transition from technology research to commercial product development is known in the industry as "The Valley of Death." Few technologies make it through alive.

While some failure is to be expected, or even welcomed, the rate is simply too high. This can lead to entrepreneurs and product managers becoming more timid; private funding requiring more and more "de-risking" before investing; and, ultimately, less innovation making its way into the hands of caregivers. The result of this high failure rate is that the founding of new medical technology companies in the United States has dropped nearly 70% in the past 30 years.[3] As an industry, this slows our innovation rate to a trickle.

This is known. It's unfortunate, but not exactly news. Medical technology development is a notoriously difficult proposition. Where things get interesting is when the discussion turns to the next natural question—*Why are these companies failing?*

Most seasoned medtech developers have theories. With our industry's high failure rate, deep and persistent introspection is certainly warranted. I have studied this for years, both academically and professionally, and I certainly have thoughts. But this is not an easy problem to solve. We must approach it as a community. And while I have theories, my personal conclusions are the derivative of many, many other perspectives whose experiences often

---

2   Medtech Engine. (2018). *Why New Medical Devices Don't Make It to Market or Fail.*

3   Innovation Counselors, LLC. (2016). *A Future at Risk: Economic Performance, Entrepreneurship, and Venture Capital in the US Medical Technology Sector.* Minneapolis: AdvaMed Accel.

resonate with my own. While every experience and situation is different, I believe that *themes* have emerged.

More on those themes in a minute. First I'd like you to hear some of these experiences from three individuals who have certainly found success in our industry, but have also been in the trenches on start-ups that failed to reach the market. We can certainly see them as mentors; as it's been said, a mentor is someone that allows you to use their hindsight as your foresight.

INTERVIEW

Dr. Robert Langer

Dr. Robert Langer is an intriguing man to research. His accomplishments are best described as jaw-dropping. Here's a smattering: he has 36 honorary doctorates; he has more than 1,400 issued and pending patents; and he is the most cited engineer in history with more than 352,000 citations. He is one of three living individuals ever to receive both the National Medal of Science and the National Medal of Technology and Innovation. In 2002, he won the Charles Stark Draper Prize, considered the equivalent of the Nobel Prize for engineers. As of this writing, he's on the *Forbes* Billionaires List as one of the top 2,000 richest people in the world. If you look him up (and I suggest you do), you'll see a smiling professor receiving awards from Barack Obama, George W. Bush, and the Queen of England. It is estimated that the products coming out of his lab have the potential to impact 4.7 billion lives. While he is called the "Edison of Medicine," one could argue that he has surpassed that moniker—and Edison himself.

The funny part is that despite this, he is still very much the engaging professor, willing and eager to help. Throughout my career I have occasionally connected with Robert, and every time I have emailed him, he has personally responded within 10 minutes. This is unheard of for someone as successful as he is.

But I'm not with Robert today to focus on his successes. Instead, I'm with him focusing on an area he is asked less about: the companies that licensed his technologies and were *unsuccessful*. Companies that he was involved with that he had to watch struggle and eventually fail. "I could name lots of them," he tells me. "Ways of delivering molecules through the skin using ultrasound, nanoparticles that could be used for the treatment of cancer and other diseases, and many, many others." Robert looks thoughtful, his hands steepled in front of his face. "On the positive side," Robert continues, "we've had many make it through the process. We've had hits, seven or eight unicorns, including Moderna. But it's sad to think about these great technologies that didn't make it through. And unfortunately, it's typically due to the company itself." Robert reiterates the point. "The technology itself is typically sound, as it's coming out of our lab, and it's our job to have those bases covered. Where we tend to screw up is on the company side."

Robert and I speak at length about how these companies fail during development, and what advice he would give someone developing a technology. "Your company needs to have a clear vision and a clear plan," he tells me. "You need to think about all the hard questions you're going to be asked and figure out how you're going to respond. A lot of time and money gets wasted when companies are unfocused. You have to have a great CEO. No C-players, no B-players. A-players only with a fire in their belly. They must be able to hire good people, have a good board, and be able to manage it. And most importantly, they must be able to communicate that vision, that plan, well enough to get investors excited."

Our conversation then turns to CEOs and investing in medtech, and the importance of one's track record. "One of the biggest challenges for medtech is raising money, but I don't think it's as tied to the whiffs as it is to the hits. The investors I know would rather take the bigger risks and rewards and focus on technologies that can be transformative. So, some failure is expected, but you just don't get that many shots on goal. If you screw up once, twice in a medical area, and the cost is too high, you're done."

As Robert and I wrap up our discussion, he strikes an optimistic tone. "While there are plenty of failed companies to be sad about, there are also plenty of technologies that are being developed to get excited about. But the area that really gets me excited is what hasn't happened yet, what discoveries are yet to be made. These can really transform medicine, and we don't yet know their names."

INTERVIEW

Dr. Susan Alpert

Dr. Susan Alpert is considered among the top U.S. regulatory experts for medical devices. She was the Director of the Office of Device Evaluation/CDRH at the FDA, Senior Vice President and Chief Regulatory Officer of Medtronic, and Vice President of Regulatory at C. R. Bard. Susan has both a PhD and an MD, so I like to think of her as Dr. Dr. Susan. I have interacted with Susan for many years, and she is always someone I look forward to speaking with. Susan is quite funny, but she is also a clear and thoughtful speaker. Where I use filler terms like "um" and "you know," Susan has none. Susan is practiced in explaining complicated concepts; each statement is well crafted and filled with information. You always leave the conversation charmed and confident that she knows what she's talking about.

As Susan and I chat a bit about some of the projects we're both affiliated with, and our work with start-ups, I ask for her impression on a phenomenon that I've noticed. I tell her that I've worked with many medical device companies, some large and some small, and many seem to be working toward FDA approval rather than working toward proving to themselves that their device is safe and effective. Susan is laughing by the time I finish my sentence.

"I know exactly what you mean," she tells me. "I'm laughing because it's something I've said to many, many people. You're not designing your device for the FDA. You're designing your device for the user and patients. The FDA is just a stop along the way. It's your job to do the design, the thinking, the testing, to demonstrate to the user and the patient what the product does. Yes, there's a regulator. But if you design a good product and test it properly, you're not going to have any trouble with the FDA."

Susan and I commiserate about how companies seem to spend a lot of time trying to guess what the FDA will think.

"It's not difficult," says Susan. "The FDA is filled with 'us.' They are the same people as we are, just wearing a different hat. There's no mystery there. When I was at the FDA, the question we would always come back to was 'What if I or someone in my family was the patient? If it was my mother, my daughter, my spouse, what do I need to know to ensure I feel comfortable with this device?'"

I have heard this sentiment repeated by many in the industry, a common gauge used when considering if what you've done is enough.

"When developing your product," Susan tells me, "you have to put on the hat as if you were at the FDA. Say to yourself, 'If I were evaluating this, not if I'm trying to market it, but trying to evaluate it, what would I want to see?' If you do that, you're going to be in the right place."

I ask Susan if she has more advice for someone going through this process. "I often remind people," she says, "that you can't assume the FDA knows anything about your product. You have to tell them anything they need to know. They are not allowed to guess. They're not sitting in your meetings. They don't have the background in your product that you have. What they have is what you gave them. You need to give them all the pieces they need to reach the same conclusion that you reached."

INTERVIEW

Dr. Manny Villafaña

It's a beautiful summer day, at least so far. It'll be uncomfortably hot later, but in mid-morning, it's perfect. I'm in an upscale restaurant in an upscale part of the Twin Cities, waiting for Manny Villafaña. When our appointed meeting time comes and goes, I decide to text Manny to tell him where he can find our table. I instantly receive a call from him asking if we were supposed to meet. Manny had removed an account on his phone and lost most of his meetings. I ask if he'd prefer to meet another time. "I'll be there in seven minutes," he says, then hangs up. So, I'm sitting and waiting for Manny, and that's just fine. The reverse would be unthinkable.

Manny is an absolute legend in medtech. Born in the Bronx in New York to poor Puerto Rican immigrants, Manny came to Minnesota when at age 27 he was hired away from his medical device sales job in the Bronx by Earl Bakken, the CEO and cofounder of Medtronic. While working in sales for Medtronic in 1971, Manny discovered a far superior battery technology than what was being used by Medtronic in its pacemakers. After Medtronic, Manny helped start a company named Cardiac Pacemakers, Inc. (CPI) that utilized the new battery technology, creating a longer-lasting pacemaker. CPI's pacemaker was guaranteed to last six years, with the industry standard closer to 12 to 18 months. CPI grew quickly and was acquired by Eli Lilly for $127 million, was spun out as Guidant, and then purchased by Boston Scientific for $27.2 billion.

After CPI, Villafaña decided to start a heart valve company, which he named St. Jude. St. Jude was eventually purchased by Abbott for $30 billion and remains a powerhouse in the industry to this day, in direct competition with Medtronic and Boston Scientific. Manny then went on to start ATS Medical, another heart valve company, which was purchased by Medtronic for $400 million.

As you drive around the Twin Cities metro area, you can't help but see Manny's impact on the medical technology industry. Dozens of sprawling, futuristic buildings dot the landscape, many of which Manny had a direct or indirect hand in creating. He has won countless awards, including the Living Legend of Medicine award from the World Society of Cardiothoracic Surgeons and the Ellis Island Medal of Honor; he is also on the Bronx Walk of Fame, and is often called "The Cardiac Kahuna." You can find Manny these days at his latest and possibly most ambitious start-up yet, Medical 21.

Manny comes in, right when he said he would, and apologetically shakes my hand. He sits down, orders coffee and toast, asking the server to not just put butter on his toast, but *slather* it in butter. "Don't worry about the heart," he tells her. "That's my job."

THE CORE CHAIN

After some small talk, we start digging into what trips up medtech developers early. "The way I see it, the most common problem early is that they didn't ask themselves the most basic question." Manny lightly pounds the table with each word. "Is. There. A. Need?" Manny pauses for emphasis. "You've got a new way to take a look at the heart? Really? Come on." Manny's New York accent flares at this. "There are ten thousand imaging systems in the world. So what? Just remember that when someone looks at your technology, if they say 'So?' you can forget about raising money. The instant reaction needs to be 'Wow.' So, need is the first thing. The other thing I ask entrepreneurs is if they truly understand how difficult this is. I'll ask them to look under their shirt or blouse. If there isn't a big red 'S' under there, you're absolutely crazy, doomed to failure. If you're going to make a medical device you've got to be Superman or Superwoman. The path is going to kill you unless you know the bullets will bounce off you. Whatever you think this is going to take in time and money, double or triple that. If you don't think you're Superman or Superwoman, go home, because developing a medical device is incredibly difficult."

12

THE POSTMORTEM

In medtech development, we all have our battle scars. When you're early in your career, you tend to sheepishly hide them. When you are seasoned in medtech development, you tend to share your scars with great flourish, like an old pirate dropping their wooden leg on the table, begging for someone to ask them the story behind it.

At the time of this writing, I'm an independent consultant. I work with as many as four or five small to large medtech companies at a time, all of which are trying to bring a technology to market. You are rarely brought in when things are easy and the path is clear. It is my job to step in, get up to speed very quickly, pick up some of the load, and help find the best way forward. I'd say about half the time, other consultants are brought in to help as well, and in the small world that is medtech consulting, you see the same people helping again and again. After a few times in foxholes together, you often develop a kinship, and before you know it, you have a core group of trusted consulting peers.

Even when we're not working on a project together, we end up talking often. We tell war stories, seek advice from each other, and even complain from time to time. When we do complain to each other, we don't really commiserate about companies, people, or systems. What really gets us frustrated, what is truly hard to be witness to, again and again, is when technologies fail to make it to market for entirely avoidable reasons.

As a medtech consultant, you are fortunate enough to be involved in the *bleeding edge* (there's probably a better way to put that) of medical technology, which is a thrilling place to be. You buy into the mission, you believe in the technology, and you get excited about the possibilities. The flip side of that, unfortunately, is the knowledge that most of these technologies will never make it to market, and you must bear witness to countless technologies never reaching the hands of caregivers.

As a result, seasoned medtech consultants get skilled at the postmortem. As we sit and share our stories, we dissect the company, the series of events, and the technology itself to debate where things veered off track. Sometimes it's quite obvious where things went astray—you knew it from your first conversation with your client. Other times the reason is more subtle, only revealing itself after long debate from multiple perspectives. We do, however, always work toward a satisfactory answer. We *need* to understand. We're often brought in to help move these companies toward success, so we strive to recognize not only what's required to succeed, but also how to avoid the mistakes others have made.

In my circle, we often frame company and product failures in one of three categories: **fortuitous events**, **technology advancement**, and **the avoidables**.

Fortuitous event failures are generally out of the company's control and could not have been easily predicted. A market crash when they needed funding, an FDA shakeup when they needed approval, a CEO winning the lottery and no longer giving a rip about the technology anymore. These events often come down to unfortunate timing and are considered part of the risk of starting a company. While it's a disappointment that the progress stops, there is little to do about it, so we learn what we can and move on.

Technology advancement failures are company and product failures that come from pushing the horizon of medical technology into uncharted waters. There are some aspects of these technologies you just can't understand until you progress them through development. Will the laboratory results translate in a clinical setting? Will the animal study results be replicated in humans? Can the technological risks be overcome? We simply won't know the answers to these questions until we try. And if we fail, these types of failures are by far the most palatable. We expect, even *want* these types of failures to occur. If, as an industry, we aren't seeing these types of failures, then we know we aren't pushing the technology far enough forward.

The avoidables are the company failures that are generally unnecessary and are wholly preventable. The company addressed a nonexistent need, they never understood the clinical workflow, the product testing was subpar, the company didn't appreciate the regulatory requirements, the company underestimated the amount of work needed, and so on. Companies don't like to think of themselves as being in this category, but if they do indeed fail, most of the time this is where the reason lies. That's why people like me tend to sit together, hashing over the avoidable mistakes that these companies make, and reminiscing about the technologies that could have been.

When we're in the trenches, we do our best to share our experiences and the experiences of other consultants, and shore up weaknesses as we see them. The problem is, however, that while individual issues and missteps can be prevented and fixed on the fly, much of what we battle on a day-to-day basis is more fundamental. What it seems to come down to, more often than not, are deep issues with clarity and alignment.

13

PART I: RECAP

KEY CONCEPTS

In Part I, we launched a new medical device company using the traditional waterfall method of product development.

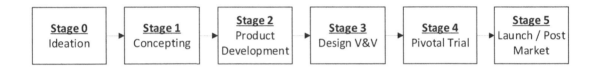

The Waterfall Method of Medtech Development

Going through the process, we were able to review many of the individual activities required for medical device product development, while establishing the waterfall method of organizing these activities. We also discussed the fact that the failure rate of medtech start-ups is high, and usually comes down to one of three overall categories:

Fortuitous Event—Failures that are generally out of the company's control and could not have been easily predicted.

Technology Advancement—Company and product failures that come from pushing the horizon of medical technology into uncharted waters.

The Avoidables—Company failures that are generally unnecessary and are wholly (but not easily) preventable. This is the largest category, and includes items like nonexistent needs, not fitting the clinical workflow, underestimating the amount of work required, not planning for reimbursement, and so on.

In **Part II** we will change our perspective from the inventor to the patient, and reconsider the product from their point of view. We will use this perspective to construct a new framework with which to view the patient and our product, along with what we're ultimately delivering.

PART II

THE CORE CHAIN FRAMEWORK

Part I	Part II	Part III	Part IV
Developing a Medical Device Using Traditional Techniques	**The Core Chain Framework**	The Core Values	The Core Chain Method of Product Development

In Part II we will switch gears, introducing a new way of thinking about medical device development, using the perspectives of our patients, caregivers, and regulators.

14

THE POINT OF WHAT WE'RE DOING

LET'S STEP BACK for a minute, if you don't mind. I'd like you to meet a friend of mine, Kyle.

(Kyle)

Kyle is a generally healthy guy. He is in his mid-30s and is an avid soccer player. Eighteen months ago, Kyle had an injury that required major surgery on his knee. The surgery went well, and after some intensive physical therapy, he regained full mobility. The pain was very severe before and after surgery, and while it has decreased with time, it remains quite painful

THE CORE CHAIN

to this day. The area is also quite sensitive and can grow red from time to time. The pain has become a central part of Kyle's existence, negatively affecting every area of his life. So, Kyle decided to see a doctor about it.

Meet Kyle's doctor.

(Kyle's Doc)

The doctor reviewed Kyle's history, asked Kyle some questions, and performed some diagnostic tests. When no obvious source of the pain was found, Kyle's doctor performed more tests. No underlying condition was forthcoming, so after deliberation and conversation with his colleagues, Kyle's doctor diagnosed Kyle with complex regional pain syndrome (CRPS).

Now, CRPS is a bit of a mystery, and doctors don't know exactly what causes it. CRPS sometimes happens when an area of the body, typically arms or legs, experiences some type of trauma. Even though the patient physically heals, the pain remains, sometimes never fully leaving. The pain can be debilitating, leaving the patient with a long-term disability as a result.

The diagnosis is bad news for Kyle, as currently there is no cure for CRPS. The best care available is treatment of the symptoms, the primary focus being pain. Kyle's doctor is frustrated too, as there are only a few things he can offer Kyle, most of which have mixed success. There are no FDA-approved drugs to treat CRPS, and secondary effects such as depression are a real concern.

While it is considered rare, Kyle is not alone. CRPS affects an estimated 200,000 people annually in the United States[4]; that's 200,000 people with a seriously diminished quality of life and no great options for treatment. So, Kyle and his caregivers are faced with a difficult battle, and no great weapons with which to fight back.

This seems like a good place to see if we can help. Let's say we want to provide Kyle, and others like Kyle, with a weapon in their battle against this debilitating disease. We are going to develop a product to help Kyle and the other 200,000ish people per year who are diagnosed with CRPS in the United States. We're going to call them our target patient population (TPP).

Target Patient Population (TPP)

This TPP will naturally be a diverse set of individuals, probably in most senses of the word. They may experience pain differently than Kyle, they may have many comorbidities (additional chronic diseases simultaneously), and they may have different tolerances and preferences to various treatment options.

We will try to help these diverse individuals by creating a solution that... does what exactly?

Well, what do they need? While the answer may seem straightforward (for example, a cure for CRPS), we must craft this carefully. If we say that the need we are attempting to address is a cure for CRPS, then we are targeting the dysfunction, which is different than targeting the symptoms, that is, pain. While we always want to fix the dysfunction, it may be nearly impossible to do so, given the current level of medical sophistication in the area. It is believed that the nerves associated with the original injury have some sort of damage (around 10% of the time), or dysfunction (around 90% of the time), causing the unrelenting pain. If

---

4 "Complex Regional Pain Syndrome," National Organization for Rare Disorders, www.rarediseases.org, accessed 9/29/2020, https://rarediseases.org/rare-diseases/reflex-sympathetic-dystrophy-syndrome

we don't really know the problem, it will be immensely difficult to create a solution. We are already at a decision point in terms of our future: we either target the dysfunction itself, in which case we'd need more research, or we target the symptoms. It seems to me that the best we can do right now is to target Kyle's symptoms, and hope that in doing so we may affect the dysfunction itself as well.

So, we are going to target the symptoms for Kyle and the target patient population. While chronic pain is the most important symptom to these patients, it's not the only one they are experiencing. Swelling, redness, and tenderness are also experienced by many people with CRPS, although it is generally agreed that the pain is the headliner. Let's focus our efforts, then, on disrupting the pain and try to affect other symptoms as a secondary or tertiary goal.

So, pain it is. Great. Well, sort of.

We need to be careful about our problem definition. Developing a solution for Kyle and the rest of the target patient population will take years of our life. There's a quote attributed to Einstein: "If given only an hour to save the world, I'd spend 55 minutes defining the problem." We're going to want to peel this onion as far as it will go before we start talking about a solution. Let's dig a little deeper.

Pain is not a binary concept. There is both a time component and a magnitude component, and each have a scale. One of the problems with CRPS is that not only does the patient experience severe pain, that pain is present for an exceedingly long period of time, often years. If we develop a product that reduces the pain for a minute, would that really help? I don't know. Ultimately, it is going to be up to the patient to decide. We know that if some pain reduction over a time period results in a substantial improvement of Kyle's quality of life, that would benefit Kyle. It makes sense to focus on quality of life rather than attempting to define exactly how much pain reduction is required.

With that, I think we have enough to make a statement that hopefully describes Kyle's need (and that of the target patient population). We will call this the need statement: *A way to address chronic pain in CRPS patients that significantly improves quality of life.*

Need Statement

A way to address chronic pain in CRPS patients that significantly improves quality of life.

With some effort, we have defined the TPP and described a need of that population we will attempt to address. If we go and talk to Kyle, asking if it would be worthwhile to develop a product that will significantly improve his quality of life, I feel confident that he would respond with an emphatic *yes*. I would bet the rest of the TPP would agree.

While we have a statement of the need we are trying to address, we don't have any real direction on how to approach it. Kyle and the TPP are unlikely to care about our design approach, so long as it's safe and it works. Our engineers do care about design approach, however, so to them our need statement is helpful but too vague. We need to further define our need statement into a list of requirements within our area of expertise.

Let's say we are going to address this need statement with a medical device (imagine that). We still have some options within that space, but we're intending to target the nerves themselves with some sort of electrical stimulation. At this point, we should talk to Kyle, Kyle's doctor, and other patients and caregivers in our TPP. If we delivered a device like that, we would need to understand what requirements or preferences they have. After getting this feedback we can do our best to distill it down to the essence of what they are requesting. These requirements become the needs of our device users, or more officially, the *user needs*:

- Device must decrease overall pain perceived by the patient
- Device must increase quality of life perceived by the patient
- Device must be implantable
- Device must be sterile
- Device must be able to be deployed in extremities and joints
- Device must not require retrieval

This list (shortened) represents our understanding of the needs of our patients and caregivers in the context of our area of expertise. If we can achieve these goals, and have a strong safety profile, our product should be successful in the eyes of the user.

THE CORE CHAIN

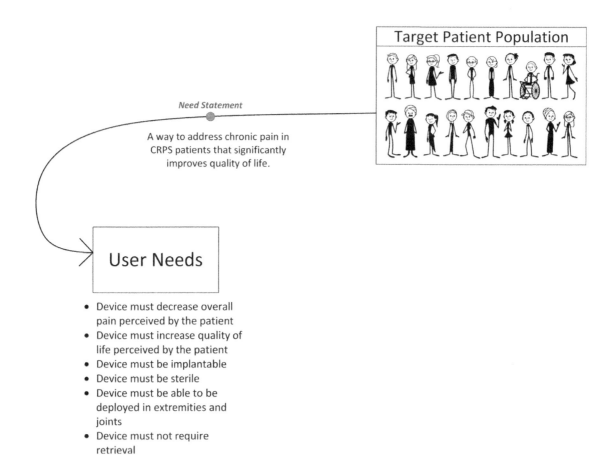

The user needs are not *elements* of the product we are making; they are *requirements* of the product. How we achieve these user needs is up to us, so long as we achieve them. The user needs are also called the "Whats," that is, *What the product needs to accomplish*. Note that we still haven't really talked about the product itself yet. That is intentional. To Kyle and others in his TPP, the product is separate from the outcome. They don't care if it's magical duct tape or a tiny bobblehead of Mikhail Gorbachev, so long as it achieves those user needs. Our engineers do care, however, as they need to create something that does achieve those user needs, and they don't think ol' Mikhail is up to the task.

So, what product will fulfill those user needs? This is what we must define. If these user needs are the "Whats," we must now create the "Hows." We need to define a list of attributes of an actual product that will fulfill what we need it to fulfill. This is no small task, and as demonstrated in the product development process, converting those user needs into a product can take years. Once this Herculean effort is completed, and we think we have developed a product that will fulfill the user needs, we must clearly describe it. The description will include everything that makes up and describes this product, including all materials, physical dimensions, surface finishes, components, software, and so on. The totality of these descriptions is known as the *product specifications*.

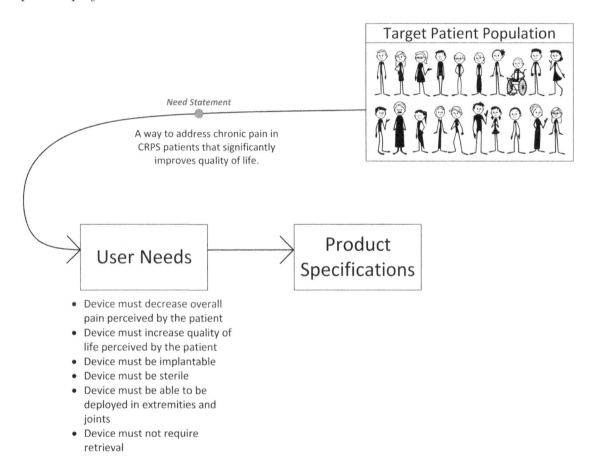

Let's say we came up with a product that we think is going to help Kyle and those with the same issue. It's a little complex, so stay with me. We have a small electronic circuit on a material about the size of a postage stamp. This little circuit (or multiples of them) can be implanted into the human body, at the site of Kyle's pain. After it's implanted, if we apply a small electromagnetic field (think Wi-Fi or ultrasound waves), it will harvest that energy and convert it into a small electrical pulse that will then stimulate the area. We can activate these circuits as many times, or as few times, as needed, and across as many of these little postage stamps as required. So, to lessen pain, Kyle can place a little mobile phonelike device up to his knee, and these little postage stamps inside his knee will start stimulating the nerves, lessening his pain. If you know what a TENS device is (little pads on the surface of your skin that make your muscle contract), this would be an internal wireless version of it. And here's the kicker: these little postage-stamp circuits will be made of a bioresorbable material that will eventually, after years of use, harmlessly dissolve into the body. Let's call this product the *fancy postage stamp*.

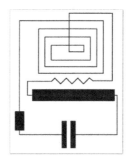

Fancy Postage Stamp

Now, we intend for this fancy postage stamp, along with a wireless energy transmitter, to be a new tool for Kyle and his doctor to use against CRPS. But we haven't really talked about a product; we've talked about an idea. We need to fully describe the actual product, in its physical form. The fancy postage stamp is made up of many materials in a very specific arrangement with allowances for real-world variation. We must describe these materials and their arrangement in detail, thereby making up the product specifications. For instance, here's how a *section* of the product specifications for the fancy postage stamp might be written:

- Polylactic acid nanomembrane base material, 30 nm thick with 3 nm +/– 1 nm thick layers
  - Overall dimension, 7 mm (+/– 0.5 mm) × 14 mm (+/– 0.5 mm) × 30 nm (+/– 10 nm)
- Magnesium (99.99% minimum purity) printed receiver circuit with 0.05 mm +/– 0.01 mm line width and 5 nm +/– 1 nm thickness
- Circuitry with the layout shown in drawing DM1201 Revision 3
- Components
  - Printed Mg/Mg0 capacitor
  - Printed Si/Mg0/Mg MOSFET transistor
  - Printed Si diode
- Printed Mg resistor

Each one of these items (and many, many more) are required for the device to function properly. These product specifications will span multiple documents, including engineering drawings. The sum of all these statements, specifications, drawings, and anything else we use to describe the device are our fancy postage stamp product specifications.

14 – THE POINT OF WHAT WE'RE DOING

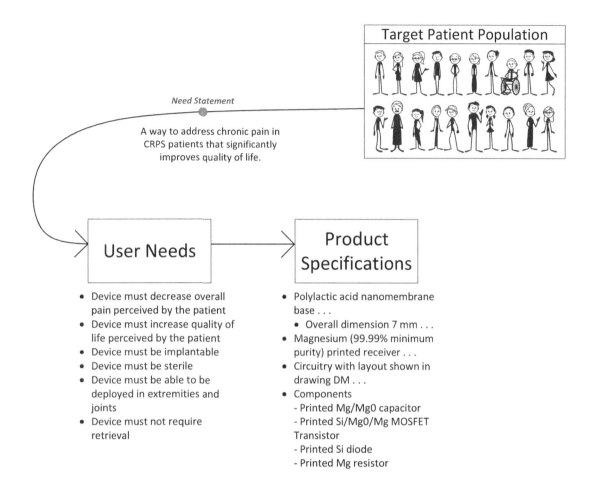

The product specifications are a big deal for everyone involved. They describe our fancy postage stamp, so we know exactly what it is that we will be delivering to Kyle and the target patient population. But describing this wonderful product and actually making it are two very different things, separated by a chasm filled with engineering hours.

15

MAKING IT

Let's recap using a simple example. If I was hungry and needed something to eat, "hungry" could be my need statement. You, as a baker, would translate that into a more detailed list of needs and preferences that you'll fulfill: food must be edible, must satiate a family of three for one meal, must taste good, must look appetizing, and so on. This would be your interpretation of my user needs. As a baker, you could say that your product will be a loaf of white bread. You would list the ingredients and attributes of the bread, maybe 10 cups flour, 4 teaspoons yeast, 3 tablespoons butter, water, salt. Maybe a bread size of 8½ inches × 4½ inches × 5 inches, light brown outside with white inside, air pockets present, compresses under light pressure, and so on. These are the product specifications, and they describe what you, the baker, are producing.

THE CORE CHAIN

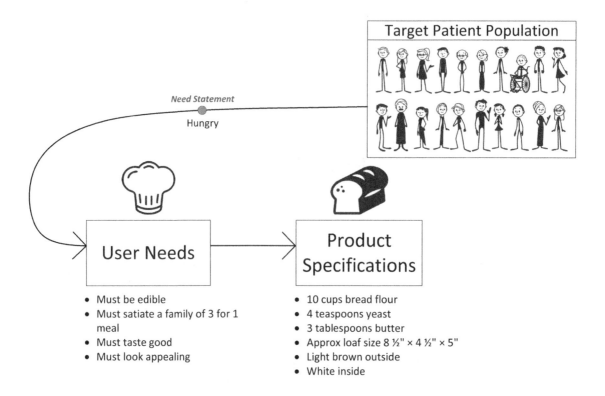

The next step is to describe how to actually make that delicious bread you defined. You need to craft a recipe, defining the steps and equipment you'll need to create the bread, along with checks you'll perform along the way. After trial and error in the kitchen, here's your recipe:

- Step 1: Combine ingredients in a mixer at medium speed for 45 seconds +/− 10 seconds
- Step 2: Knead for 8 minutes +/− 1 minute on a surface powdered with flour
- Step 3: Place dough ball into a greased bowl, cover and let sit for 1 hour
- Step 4 (Check): Verify that the dough has approximately doubled in size. If not, allow additional time to rise
- Step 5: Put the dough into a greased bread pan, and bake in oven at 400 °F +/− 20 °F for 33 minutes +/− 3 minutes

This is the detailed recipe for the bread, also known as your *manufacturing specifications*. If you follow this recipe correctly, you should end up with the bread you defined in your product specifications. If the recipe is well defined, you should be able to execute it hundreds or thousands of times and get the same result: delicious bread for hungry people.

15—MAKING IT

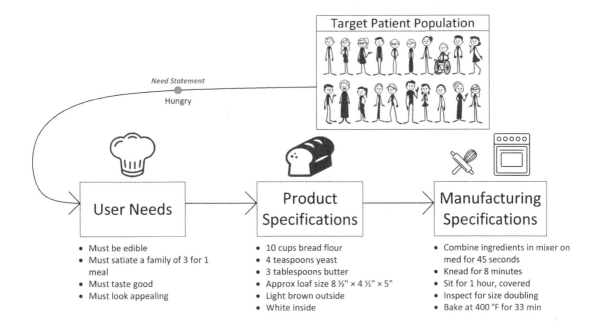

This, then, brings us to the same point that we are at with our fancy postage stamp. To address the TPP, we will need hundreds of thousands of fancy postage stamps, so we'll need to create a recipe that will result in the product we defined.

To manufacture our postage stamp, we will take the components defined in the product specifications and apply processes to them to create our product, same as the bread. The manufacturing specifications can be in step-by-step order, starting as far back as we can, ideally all the way from raw materials. Here's how ours might look for the fancy postage stamp:

- Base substrate material construction
  - Dissolve polylactic acid granules in acetone to a 10% w/v concentration
  - Magnetic stir at 500 rpm at room temperature for 4 hours
    - Equipment used: analytical balance, magnetic stirrer
  - The polylactic acid is spun into fibers using electrospinning equipment with the following parameters . . .
    - Equipment used: 45 kV power supply, rotating cylinder, grounded collector . . .
  - The fibers are knit into a woven structure with a fiber density of 300 PPI

THE CORE CHAIN

- The knit fibers are spray coated using ultrasonic spray coating at 2000 rpm using an aqueous gelatin solution . . .
- Circuitry
  - Energy harvesting antennae, capacitors, leads, and connective circuitry are screen printed using ultra-high purity magnesium . . .

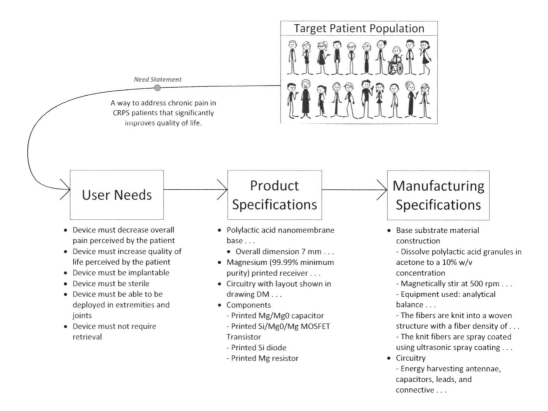

The manufacturing specifications are our recipe to follow to create our product. They should be at a level of detail where, with proper training and equipment, any operator will get the same result. This recipe, along with the product specifications (what you're making), is also known as the *device master record* (DMR).

Our last logical step is a straightforward one: we need to follow this recipe we crafted and create our product. If our recipe is good, then it should be straightforward to follow, without much room for interpretation. When someone executes that recipe, the result is the *batch record*.

15—MAKING IT

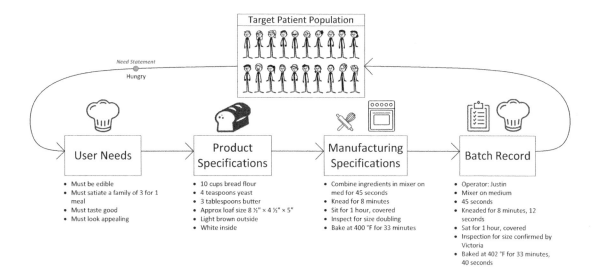

The batch record shows the actual values that were used when executing the recipe. The weighed-out flour, the actual temperature of the oven, the people executing the recipe, and many more items are captured in the batch record. This record is only applicable to one batch of bread made; a new one will need to be created each time the recipe is executed.

The loaves of bread are the physical output of this process, delivered to me and others like me, to address the need that we have so emphatically expressed. The same is the case for Kyle, his doctor, and the target patient population. When the manufacturing specifications for the fancy postage stamp are executed, you create the batch record, which tells the story of making the batch, the operators who performed the activities, the amounts of each raw material used, the equipment used along with their settings, and the results of the various product inspections. For instance, a batch record for the fancy postage stamp may contain records like this:

- Base substrate material construction
  - Polylactic acid prep operator: Chris
    - Volume of polylactic acid: 100 g
    - Volume of acetone: 900 ml
    - Magnetic stirring time: 4 hours, 22 minutes, 6 seconds
  - Electrospinning operator: Linda
    - Power setting: 45.0 kV
    - Cylinder rotation: 90 rpm
    - Tensile strength: 1.1 mpa
  - Ultrasonic spray coating operator: Julius

THE CORE CHAIN

- Ultrasonic frequency: 95 kHz
- Air pressure: 12 psi
- Time: 7.3 seconds side 1; 7.2 seconds side 2
- Final deposition weight: 34.8 mg

As this results in the finished product, the batch record is the last major element of the logical process, from an expressed need of a patient to the fulfillment of that need.

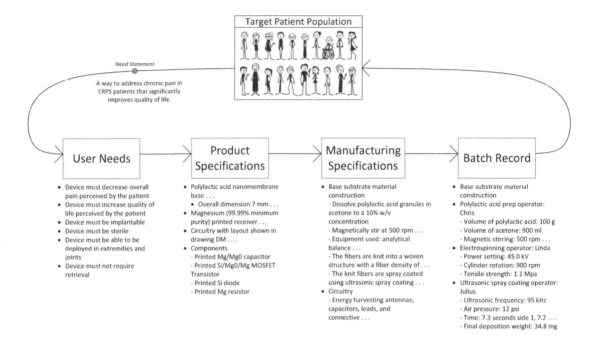

Delivery of the product into the hands of the patient and caregiver is the point of what we're doing. Kyle and his doc need more tools in the effort to battle CRPS, and we are handing them one. That is the completion of the logical progression, and hopefully the fulfillment of the need for our dear friend Kyle.

Now, in theory, if our batch record shows that we executed the manufacturing specifications correctly, that should mean that our product is good, right? Well, do we really know that? At this point, we actually don't.

Kyle is counting on us to get this right, so let's make sure this is correct before we ask his doc to stick these into his knee.

16

TYING IT ALL TOGETHER

In God we trust, all others must bring data. —W. E. Deming

At the beginning of our little expedition, we talked to Kyle, his doc, and others in the target patient population (TPP) to formulate what they need from us, that is, the need statement.

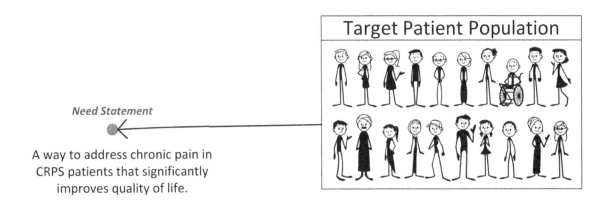

The need statement is never a nice, clear, consistent message provided by everyone in and around the target patient population. Instead, it's *distilled* from what we heard, what we understand about the disease, and what we understand from the experience of affected patients. The need statement is our words, not theirs. What if we were wrong? Well, that would be bad, because we based all our work on fulfilling that need statement.

THE CORE CHAIN

Before we go too far, we need to ensure that the need statement nails the need. This is where a validation activity comes in. A validation activity is intended to prove the validity of something (imagine that), like our need statement. To ensure something is scientifically valid, we need objective evidence that demonstrates its validity. In the case of the need statement, we need to perform a need statement validation by asking individuals in our patient population, as well as caregivers and experts in the area, to challenge the statement *objectively* and ensure that it is a solid representation of the need. Once we do that, we can have confidence that we are addressing a real and urgent need.

So, we derived the need statement from our patients and caregivers. We then must test that need statement *back against* that patient population. We'll call it our need statement verification.

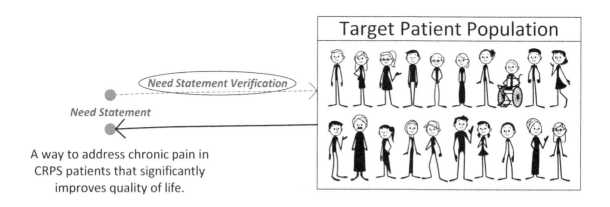

Let's pause here and take a minute to visualize what we're trying to do. As we built the core chain, we started with the target patient population, did our best to understand them, then built a bridge from them to our user needs. This bridge connects our patients to a set of user needs that we will use moving forward. We then built bridges connecting us to our next stop on the chain, the product specifications. But just because we built these bridges doesn't mean they're *good* bridges. And, as a general rule, if I'm unsure if a bridge is good, I'm not going to use it. This is the point of our validation and verification activities—going back, in a methodical and serious way, and checking our bridges. Also, I wouldn't think of these as sweet stone arches over a bubbling brook; I'd think of them as attaching two distant cliff faces together. The drop is . . . notable. It is in our best interest (and the interest of the patients) to make sure our bridges are solid. How we do that is dependent on the bridge, but these verification and validation activities all try to accomplish the same thing: checking if they were built correctly.

16—TYING IT ALL TOGETHER

The need statement verification activity tested the bridge between our need statement and target patient population. Next, we'll want to test the bridges between our all-important user needs and the target patient population that they are intended to represent. This bridge check is called *needs validation*.

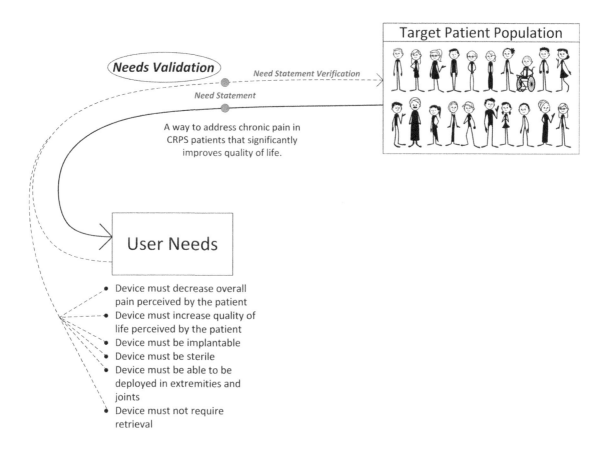

I will note that, in my experience, these bridges are often the weakest in the core chain. This is concerning, as the user needs are what we design our product to fulfill, so we need high confidence that they are correct. The needs validation activity is then critical, as it challenges one of the trickiest and most important set of bridges in our core chain.

If we are able to validate the links between the user needs and the target patient population, we can have confidence in our user needs. This gives us confidence in building the bridges to our product specifications. It's probably important to note the inverse of this statement here. If the

THE CORE CHAIN

user needs are not known to be strong, there is little to gain by building product specifications based on them. That seems obvious, but in practice, it isn't.

This testing of the bridges between product specifications and the user needs is known as *design validation*.

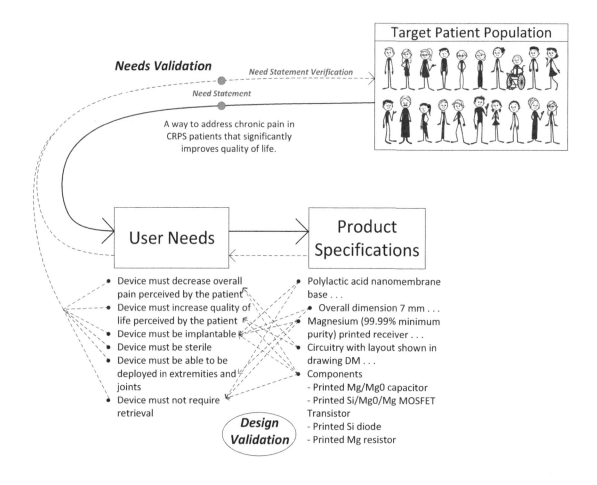

These bridges between user needs and product specifications are big ones. Really, really big. So big, in fact, that we as an industry add a middle step between these two to ease the transition. Rather than trying to design straight from the user needs, an additional set of requirements is added. These requirements are known as the *design inputs*.

Think of the design inputs as a large support added in the middle of this large bridge between the user needs and the product specifications. These design inputs restate the user needs into more manageable and testable statements that we can design our product to and test against.

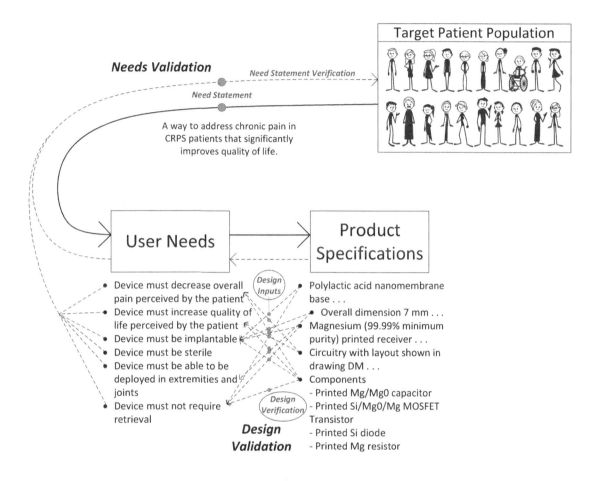

When design inputs are crafted well, if we fulfill them through the product specifications, we also fulfill the user needs. If a User Need is "Device must be implantable," then a design input may be something like this: "Device must be implantable behind the superior extensor retinaculum in 95% of male and female adults." This provides a much more specific target, and it is helpful, so long as it correctly clarifies the User Need.

We first test the section of the bridges between the design inputs and the product specifications. This is called *design verification*. When we're sure those sections are strong, we test the overall bridges between user needs and product specifications through design validation, which often includes a clinical trial. Between these two activities, we can gain confidence in our bridges between the user needs and product specifications.

One note here: we don't necessarily test all these bridges with the same rigor. For each bridge, we need to understand the impact if it's unsound. We do this through the risk analysis process. Basically, how far we (or the patient) would fall if that bridge broke. I spoke earlier about these bridges connecting two cliff faces. While that's true for some (or most) of these bridges, there are some that are much closer to the ground with little impact to us or the patient if they break. We change the rigor of our tests based on the impact (or severity) of failure. For the bridges between the user needs and the product specifications, we typically assess impact through the design and use risk assessments.

The next set of bridges in front of us are those that connect our product (product specifications) with our recipe (manufacturing specifications). The testing of these bridges is known as *process validation*.

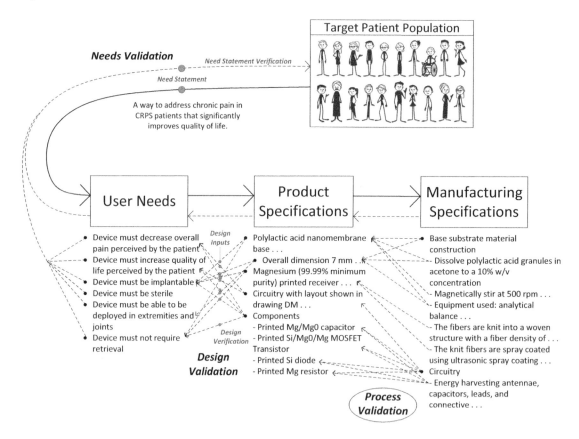

16—TYING IT ALL TOGETHER

These tests are intended to verify that the recipe we defined consistently results in the product we specified.

If one of our processes is curing the final device in an oven, and the device strength changes during that process, the temperature and time in that oven could affect the overall strength of the product. We can then link the "oven temp and time" manufacturing specifications to the "strength" product specifications. Let's say that through testing we know that if the oven temp is 380 °F–420 °F for 15–17 minutes, we end up with the correct product strength. The process validation shows that if we stay within those ranges, we will make product that meets the product specification for strength. We need to do this for all elements of our process that can affect our product specifications (unless we decide to inspect 100% of our product). After you are done validating each individual process, you perform an overall validation of the manufacturing, running through the entire manufacturing process multiple times, testing the finished product.

With process validation, we ensure that our processes, when run within allowable ranges, result in a product that meets our product specifications. That's important to know, but when actually performing the manufacturing, how do you know that the manufacturing team will run the process within those allowable ranges? That's a fair question, and it brings us to the testing of our next set of bridges, between the manufacturing specifications and the manufacturing batches.

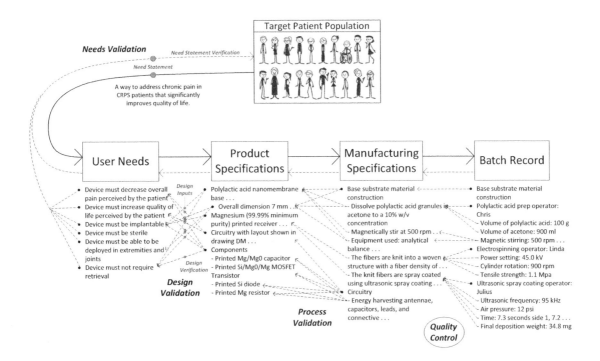

THE CORE CHAIN

The connection between the batch record and the manufacturing specifications is tested by quality control. As building batches is an ongoing activity, quality control (QC) performs their checks on the bridges every time a batch is built. QC checks that when the batch is executed, the recipe was followed properly. They check the settings, test the product, confirm that people were properly trained, equipment was properly maintained, and so on.

The last bridge we need to check is between the delivered product and the users for which it is intended. This verification is arguably the most important one of all, as it provides direct feedback on our product from the users and patients themselves. This is aptly named *user feedback*.

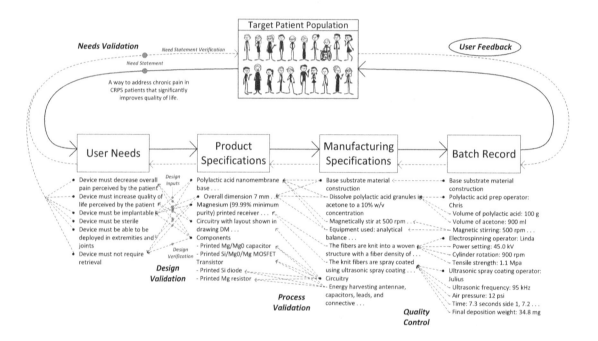

The user feedback tells us whether the product we delivered to the users is performing as intended and within expected safety and efficacy parameters. These checks, performed graciously by the user, are a little unique. They not only test that last bridge from produced product to the hands of the user, they also provide a check on all previous bridges we made as well. If we get feedback from the user that the product doesn't fit as intended, then we need to check our product specifications, as well as the bridges we validated between them and the user needs.

16—TYING IT ALL TOGETHER

This completes the loop or, as I like to call it, the chain, from our patient population need to feedback on whether we actually fulfilled that need with our product.

If this all looks like a lot of work, frankly, you're right. It's hard enough to develop a product without having to do all these validation and verification steps. Do we really need to be this thorough? Well, let's return to our friend Kyle.

(Kyle)

Kyle has been patiently waiting for us to create a product to help him. When we finally deliver a product to him, that product may fulfill Kyle's need. It's also possible that it won't do any good at all for Kyle. In the worst case, it may actually severely harm him. That is a tremendous responsibility. How does Kyle's doc know that when we hand him the product, it will perform as intended for Kyle?

We know that the product we provide for Kyle and his doctor is safe and effective through that chain of confidence we've built, which I like to call *the core chain*.

THE CORE CHAIN

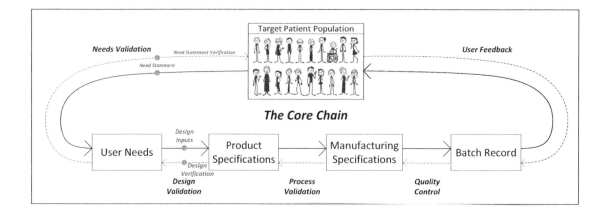

The core chain is the logical progression that ensures that when we deliver that product to Kyle, and anybody else in the TPP, it performs as we intend it to perform. It's what we spend the vast majority of the product development process building, and it's what the FDA evaluates when they look at our technology.

We can use this core chain to always trace our activities back to the patient. For instance, using the core chain, when we bake the product at 400 °F for 16 minutes, we follow this logical progression:

- Through *Quality Control* **we know** that the temp and time on the *Batch Record* are within the *Manufacturing Specifications.*
- Through *Process Validation* **we know** that if we are within the *Manufacturing Specifications* that we will be within the *Product Specifications.*
- Through *Design Validation* **we know** that if we are within the *Product Specifications*, we will meet the *user needs.*
- Through *user needs Validation* **we know** that if we meet the *user needs*, we are truly helping Kyle and others in the *Target Patient Population.*

The core chain represents assurance. We spend thousands of engineering hours building it before we deliver the first product. We will rely on the core chain every time we create a new batch of product, for the life of our production.

So long as the core chain was built correctly, and remains intact, you know that the product delivered to patients and caregivers will fulfill their needs. This assurance is what anybody looking at your product wants to see. If the FDA shows up for a visit (which they like to do, with a few days' notice), they will start on the far right of the chain, at user feedback (aka post-market surveillance), and slowly and methodically work their way back along the chain, making sure everything was properly built and is intact, all the way back again to the users. They will be looking for weakness not only in our core chain itself, but also in your company's ability to build and protect that chain.

The core chain provides the framework for understanding the relationship between all aspects of our product, its manufacturing, and its impact on the patient. Not only does it provide a visual understanding of these relationships, but it also forces you to reexamine the product development process itself.

17

WHAT WE'RE BUILDING

Let me ask you a question. What are we, as medtech inventors and companies, trying to develop during product development? If you answered, "a product," you'd be wrong, strange as it seems. If we developed a product without building the rest of the core chain, then started selling it as a medical device, we might get thrown in jail. A product is just one piece of what must be delivered as an overall package. The logical chain from patient to delivered product, connected by objective evidence, is what gives the product any value. So, if we're not actually developing a product, you might ask why we call it "product development." That, my friend, is a great question.

First and foremost, we need to be clear about the goal of the product development process. In medtech, "product development" is a misnomer. It points us to the product itself as the goal of our activities. All regulatory requirements (like validating the product) seem like hurdles in the way of our goal. This is, unfortunately, how many companies veer off track. We are not developing a product. We are developing confidence.

A few chapters ago, I lamented "the avoidables," the company failures that could have been prevented. I argued that it often comes down to issues with clarity and alignment. If we misunderstand the goal of our activities, an issue with clarity is to be *expected*. The core chain was created to show individuals and companies exactly what we're trying to create in medtech. When we remove the *product* as the goal and replace it with *confidence* provided by the core chain, the development mindset changes. We now have visibility into the entire package that we must deliver to patients and caregivers. We can document, visualize, and trace every piece of that complete package as we build it, see which connections are weak, and which connections are robust. All our activities fall into a more complete blueprint, and our daily discussions and decisions take on a clearer context.

THE CORE CHAIN

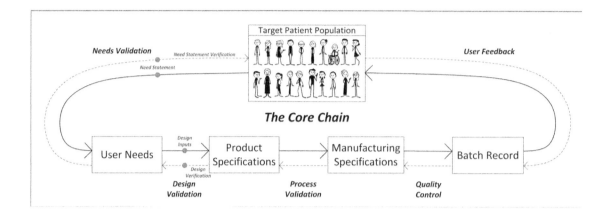

The core chain gives us a framework to understand what we are building and why. We can see how our activities contribute (or don't) to the logical chain between the end product and the patients and caregivers who need it. For me, the core chain oriented and organized my thinking around bringing a medical device product to market. As I sit in a conference room, discussing a problem that a company is facing, the core chain inevitably ends up on the board, with a big red circle around the point where the problem resides.

Also, when performing postmortems on the companies that went astray, I find it helps to examine these companies through the lens of the core chain framework. The errors that led to these failures become more concrete and even *visible*. When these errors are visible, they become even more apparent when repeated across companies, again and again. I have found that companies often make the same mistakes as their peers, as we share an industry history, regulatory landscape, and talent pool. Luckily, this means that if these mistakes are predictable, avoiding them becomes possible.

Using the context of the core chain, we can distill these repeated learnings into something of substance. As we now have clarity in what we're building, how our efforts fit together, and our industry's common mistakes, we can start to refine our approach to increase our chances of success. If medtech development is a wilderness, the core chain is our destination. But we still need a path, as well as principles to follow, to help keep us out of the quicksand.

18

PART II: RECAP

KEY CONCEPTS

In medical device development, the *product* is not the end outcome of our activities; the end outcome is *confidence,* the confidence that our medical device actually delivers the care it is intended to deliver. We represent this chain of confidence between the patients, their requirements, our product, and its manufacturing by the *core chain*.

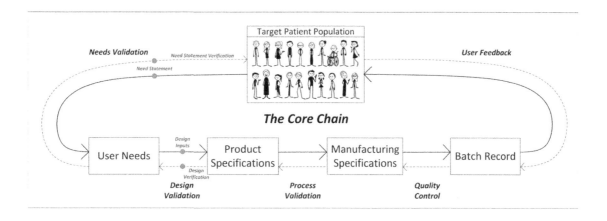

The core chain framework provides the foundation on which we can build our principles of product development, which we call *the core values*. We will describe these core values in Part III.

PART III

THE CORE VALUES

Part I	Part II	Part III	Part IV
Developing a Medical Device Using Traditional Techniques	The Core Chain Framework	The Core Values	The Core Chain Method of Product Development

In this part, with the core chain framework understood, we will cover a set of eight principles that we can follow that help us avoid common failures in the product development process.

19

CORE VALUE #1—BUILD THE RIGHT CHAIN

When I was studying for my undergraduate degree at West Virginia University, one of my professors was a fun, energetic Brit named Nigel. Nigel was a whip-smart engineer, drove old British cars, and had the accent to match (which, as someone fresh off the farm, I found fascinating). Nigel was also a perpetual inventor. During one of my all-night cramming sessions, I was chatting with another student when he mentioned that he was working on an invention with Nigel. Surprised, I demanded to know more, and he brought me to one of the automotive labs in the building. There in the lab was a highly modified engine like I had never seen before. I asked him what I was looking at, and he told me I might be looking at history, because this was the very first CIBAI engine. I asked him what the heck a CIBAI engine was, and he told me it could change the way a car engine fundamentally operates. He then went on to explain the concept, which went something like this:

The CIBAI engine functions with two cylinders in an engine working in tandem, one cylinder with an air/fuel mixture, and one with just air. When the two cylinders reach the top of their stroke, the air-only cylinder compresses the air in its cylinder to extreme pressures, at which point a valve opens between the two cylinders. The high-pressure air rushes into the other cylinder so fast that it breaks the sound barrier, creating a shockwave and a mini sonic boom, igniting the air-fuel mixture. The explosion then rushes back to the other cylinder along with the first, pushing both cylinders down. No spark plugs or high-pressure fuel system needed.

I found the CIBAI engine to be a brilliant idea, and a rare combination of the two degrees of my program, aerospace and mechanical engineering. I loved the idea of an engine that was creating internal sonic booms many times a second. Not only did it work, but it burned fuel more cleanly, resulting in something like a 10% or 20% reduction in emissions while bumping up fuel efficiency.

For weeks I often thought about the invention and its potential implications. It was exciting to be around inventors pushing the state of the art, and I wanted to be part of it. One day I was seated in the commons area of the engineering building, talking with another student about the idea. One of our other professors happened to walk by and I called out to him, "Professor!" He looked and walked over. "What do you think about Dr. Nigel's CIBAI engine idea?"

"Oh, it's absolutely brilliant," he said. "But of course it won't end up being anything."

"What? Why?" I asked, obviously surprised.

"Well," he replied, "the patent office is filled with ideas that will improve fuel efficiency by twenty percent. Making CIBAI engines would require a complete retooling of the industry. It just wouldn't be worth putting it into practice."

I stared at him for a second. "Then why are they spending so much time and money on it?"

"Dunno," he said. "Probably because it's a good idea, and they can't let it die."

Looking back, this scene is still clear to me, burned into my memory. It took a while to fully process what he said. It was the first time as an engineer that I had to mentally separate the quality of an idea from its real-world usefulness. I'm guessing that if the idea was mine, I wouldn't have had the mental fortitude to separate the two.

Working with many new inventions over the years, I have seen this scenario play out many times. Much time, effort, and resources have been spent on good ideas that don't have much real-world usefulness. Now when I'm asked to look at a company's work, I try to assess it by asking the following question: *Is this the right chain for this landscape?*

CORE VALUE #1–BUILD THE RIGHT CHAIN

There are few industries where this value is more important, as the modern healthcare system is notoriously complex. The risk/benefit equation you present to the patient is only part of a larger ecosystem. Systems, workflows, and entrenched mindsets dominate the landscape. For a product to be successful, it must not only work for patients, but also for the stakeholders that surround them. If a surgical instrument has marginally better outcomes, but costs more, and requires notable changes to the workflow or physician training, it's probably dead as disco.

UNDERSTANDING REAL-WORLD USEFULNESS

Understanding the real-world usefulness of your product starts with understanding the landscape in which it will exist. The landscape is highly dependent on the product and its clinical pathway. The landscape of a direct-to-consumer product is very different than a product intended to be used in an operating room. The larger and more entrenched the systems and workflows, the more your product needs to operate within them. Here is a list (not comprehensive) of potential landscape attributes to understand:

- Caregiver workflow—In situations where your device solution will be deployed, how will it fit into the ways caregivers currently address that solution? Does this add time or save time in the caregiver's hectic schedule?
- Administrative workflow—How does this device fit into the systems of the hospital or clinic as a whole? How will they order it, how will it enter the building, where will it be stocked, will it be pay up front or pay when used, and so on?
- Clinician training/mindset—What does the clinician currently do in this situation, and what is the standard of care? What would they personally need to change to utilize this product? How painful is that change, and what's their benefit if they do make the change?
- Payor system—How will the hospital or clinic get reimbursed for this? Does it use a standard reimbursement code, and if so, will the payor actually pay for it? Will this add to or reduce overall healthcare costs? If it reduces costs, can this be proven?
- Patient mindset—What will the patient feel about this solution? Would they have trepidation? Are there other therapies they'd likely prefer to try prior to this one?

Now, it can be said that a revolutionary product can overcome any of these items. But in that case, the benefit of your product must be so large that it outweighs the pain of change (usually by a significant margin). A CIBAI engine with a 20% efficiency bump may not make it, but a 50% efficiency bump might justify making all those system changes. Surgeons will learn a new approach if it significantly helps their patients, and payors will create new reimbursement codes if it is to the benefit of everyone involved—but you must give them very compelling reasons to do this.

UNDERSTANDING THE COMPETITIVE LANDSCAPE

The other landscape to understand when considering your chain is the competitive one. Companies tend to block off large sections of patent space for a particular approach to a problem. The available patent space may limit (sometimes dramatically) the engineering options you have to address your need statement. Every company must understand this, and many large companies in similar spaces have a constant turf war happening in a patent space. Outside of the patent space, you must consider the competitor companies that are attempting to address the same patient need. Their strength, entrenchment, level of activity within the space, and future plans must all be taken into account. The development of your sales force will take years, and you should know if there are multiple major companies innovating in your space with a head start.

The competitive dynamics will help set the limits around your core chain. For instance, when translating from a need statement to the user needs, you may want to stay out of the active

THE CORE CHAIN

implantable space, as it is crowded and the competition is fierce. Or, when creating your product specifications, your freedom-to-operate in the patent landscape may cut off a few design pathways. All of this is considered normal and should be understood as you build and rebuild your core chain.

The real-world usefulness and the amount of competition across an indication (the disease state you are targeting) should be looked at in concert. In general, in mature spaces, if the level of usefulness for a particular approach is high, so is the level of competition. As the level of usefulness goes down, so does the level of competition. An indication will have a horizon line between these two attributes, where the current best choices available to the clinicians/patients can be mapped.

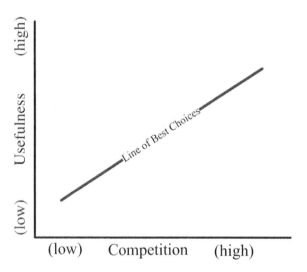

The Usefulness-Competition (UC) Chart

The line of best choices represents all the different technologies that are attempting to address the indication (for example, coronary heart disease). Some are medical devices, some are pharmaceuticals or biologics, some may even be homeopathic remedies. Each one of these technologies has a level of usefulness and a level of competition. Typically, there is a leader or two in each of these technologies, and they have the honor of living *on* the line of best choices available to the patient and caregiver.

Where your core chain lands on this chart is significant, as it greatly affects how you should approach your technology.

19—CORE VALUE #1—BUILD THE RIGHT CHAIN

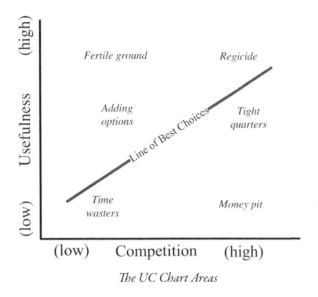

The UC Chart Areas

If a particular core chain approach you are considering sits in the "regicide" area, you are going up against massive companies in highly contested areas. You may want to consider alternative needs to address within the indication, which will move you into lower competition areas. If your core chain is sitting below the line of best choices, you need to consider how to increase its usefulness by adding features, improving the workflow, reducing the cost, and so on. In general, you are aiming to be as far above the line as possible, while considering how difficult it will be to achieve this aim in this area (from both a competition and technological perspective).

As you would expect, these UC charts are different for each indication. CRPS may have a UC chart that looks like this:

THE CORE CHAIN

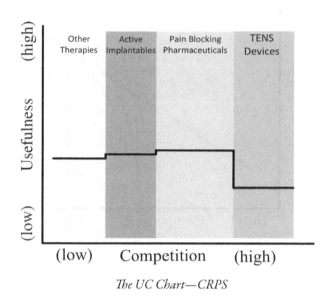

The UC Chart—CRPS

While coronary heart disease (CHD) may have a much different UC chart:

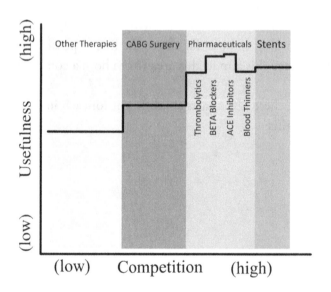

The UC Chart—Coronary Heart Disease

Creating the UC chart helps you understand the lay of the land, and where your core chain fits with patients and caregivers. It can also help you strategize your business approach. If you attempt a regicide in a well-worn indication, you will need a lot of resources at your disposal, and you can expect a fierce competitive response. If the line of best choices is generally low and the indication underserved, you may find that any usefulness increase is enthusiastically received by clinicians, patients, hospitals, and payors (and sometimes even competitors).

CORE VALUE #1—BUILD THE RIGHT CHAIN

As you create your core chain, you must constantly challenge where it fits in the UC chart, and where the line of best choices resides for the indication. If you always keep your eye on this, you will help ensure you are building the right chain.

20

CORE VALUE #2 – BUILD FROM LEFT TO RIGHT

MATT EMMONS, A 23-year-old accountant, stared down the scope of his rifle. He had dominated the 50-meter, three-position rifle events all year and was on the final shot of his final round. He was one shot away from an Olympic gold medal. Matt needed a 5.2 to win, in a sport where anything below 8 was considered amateurish.

Matt aimed carefully and squeezed the trigger. Matt smirked from behind his scope, knowing he shot an 8.5. He waited for the score to display and the crowd to erupt.

But no score popped up. He gave it another second to register, but nothing happened. A few seconds later, Matt lifted his hand off the rifle in a questioning gesture. He looked back at the officials. The officials said he didn't fire his gun. Matt tells them he did. He pulls the spent shell casing out of his chamber and shows them.

Minutes later, Matt's head is down but nodding as the official announces to the crowd that Matt's score is a 0. In an extremely rare mistake in elite competition, Matt had fired at the target in the next lane. Instead of Olympic gold, Matt ended up in eighth place.

Despite years of meticulous training, millions of practice shots, and being one of the most talented people in the sport, it all came down to the fact that Matt didn't confirm whether he was looking at the right target when he took his shot.

THE CORE CHAIN

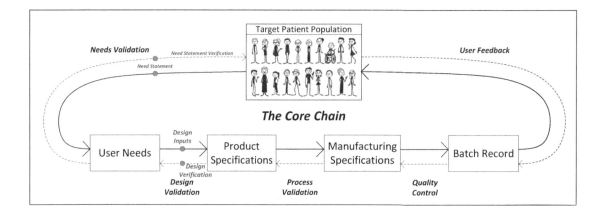

When looking at the core chain, consider the direction of the arrows. The solid arrows show the logical progression along the chain, and the dotted arrows show the perspective of each step's validation. The direction in which these arrows are pointing is significant and can tell us a lot about how these activities should be performed. When creating the user needs, they are born out of the need statement, and represent the next logical step toward a physical product. Once these user needs are created, we can validate them back against the need statement and patients from which they were created. When this relationship is continued all the way through the chain, we maintain that logical progression from one step to the next, with each element in the chain being the fulfillment of the requirement to its left, and then becoming the requirement to the element on its right.

This intuitively makes sense; it's how we ensure that the product we are handing to the patient actually fulfills the need that we were targeting. This also has some interesting implications. Each link in our chain requires the one before it to be correct, so if an early link is incorrect, all links tied to it are incorrect. However, if a late link is incorrect, for instance in the batch record, everything before it may still be intact. Fixing links on the right is much easier than fixing links on the left.

Every link in the chain relies on the one before it, so if your early links are weak (or non-existent), it doesn't matter how strong the downstream ones are, your chain is weak. The farther left on the chain, the more foundational the links become, and the more difficult they are to change as you continue down the product development pathway.

Having an idea is an exciting and admirable feat. The problem is that starting with an idea can violate the direction of the arrows shown in the core chain. By working through the idea first, you are starting with the product specifications without any requirements in place.

You then work backward along the chain to create the user needs, deciding what the product needs to accomplish after you started trying to accomplish it. Starting with the idea is ripe for various types of bias, and the user needs often take on the perspective of the inventors rather than the patient. A lot of time and effort is then spent on proving that the product is accomplishing those requirements, when it is of course destined to do just that because they were made with that product in mind after all. When the product is finally challenged by people without those same biases (patients/doctors/FDA), it often languishes or outright dies.

However, this doesn't mean that you can't start with an idea. In reality, that's how most products start. But if you do, the confirmation bias trap needs to be carefully avoided, egos be damned. An idea must be *set aside* until the links before it can be independently created and verified, then the idea can be evaluated against them.

CORE VALUE #2—BUILD FROM LEFT TO RIGHT

Companies that embrace the core value of "build from left to right" are challenged to put a much larger focus on these early links in the core chain. To build from left to right means to do three things:

1. *Focus on making the left side of the chain correct before progressing down the chain.* This does not mean you should fully complete these elements of the chain before progressing (see Chapter 25: "Rough It in and Refine"), but it often means focusing much more time and effort on the perspective of the patient and caregivers.
2. *Kill bias.* We are looking for unbiased, unvarnished truth about the needs of the patient, regardless of how it affects us. Independent creation and validation are key here.
3. *Challenge early and often.* Do not wait until late in the process to ensure you're aiming at the right target. As your core chain is built, the left side of the chain should be challenged above all others, with increasing effort as you progress.

"Building from left to right" is one of the more challenging core values to instill in a company, as it often comes down to the egos and desires of those involved. Following this core value is an attempt to get the unvarnished understanding of the patients' needs and flowing all efforts out of those needs. It is then a continual challenge to ensure the accuracy of those needs, regardless of how it affects the company.

21

CORE VALUE #3–PREVENT TRANSLATION ERRORS

Early in my career I lived in Zhuhai, China, leading a project to set up a new medical device production. The large multinational project team included a group of young interpreters who were bright, energetic, and eager to sharpen their skills.

I had one or another by my side most days, translating my requests and communications with the largely Chinese project team. For important conversations, we would often have a pre-meeting to go through the message before they needed to translate it in real time. During one of these pre-meetings, while talking about an important point that I was repeating, I said, "I don't want to beat a dead horse here." With arched eyebrows, my translator said, "What horse?"

"Sorry," I said. "There is no horse. It's an expression."

"An expression of what?" she asked.

"I guess the pointless activity of doing something that has already been done," I said, trying to explain.

"You mean beating a horse to death?" she asked.

"Um, I suppose so," I replied, starting to get uncomfortable.

Unfortunately, the conversation continued:

"Do you often beat horses in America?"

"No!"

"I know not dead ones," she said, "but do you beat live horses to death?"

"I really don't think so," I said mumbling, head face down in my hands.

"It seems like a weird expression then, why would you ever beat a dead horse?"

"You wouldn't, that's the point."

"So the point," she said, "is don't waste time beating something that you have already beaten to death."

"Sorta, I guess. Maybe it's easier if I just say I don't want to go over this too much."

THE CORE CHAIN

"Sure is easier. Poor horses."

If she had translated what I was saying live, a group of about 70 team members would have been convinced that I was brutally beating horses in my spare time. Luckily, she could sense that a translation error might be coming and took the time to understand what I was trying to say, not exactly what I was saying. History is rife with these types of flubs, and unsurprisingly so, as translating is tricky business.

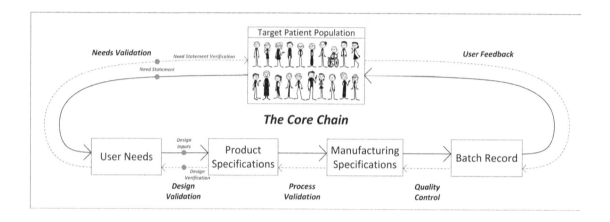

From each step in the chain to the next, one of these translations is occurring. The patients or doctors do their best to translate their pains, concerns, and desires into words, which we then do our best to translate into one concise statement of need. We then take this brief statement and translate it into a set of requirements that a product must fulfill. We then translate these requirements into the physical realm, creating a product that does what we intend it to do. Throughout these translation events, there is a set of intangible threads that we are trying to convey through various means. We need to identify and protect those threads as they converge to create the care proposition for the patient.

I was lucky to have thoughtful, careful translators on my team, and even then, confusion was a constant battle. Many others have not been so lucky, with translations performed word-for-word with little effort made to understand intent. Through impression, diction, tone, and context, the best translators have an almost supernatural ability to decode far beyond what was said, peering deep into the mind of the sender. They then craft the message into another language, while being deeply aware of their own translation biases.

We are translating not only between people and points of view, but between the tangible and the intangible, while trying to keep the message intact. Ultimately, we are translating a set of human impressions into a physical product. To get there, we must strive to move these impressions and ideas along with the same high level of care, as the stakes are often rather high for us—and the patient. Luckily, we do not need to perform these translations live. We can use time, along with a set of tools and techniques, to ensure that we are able to thoughtfully craft these ideas into the target language, whether they be statements of intent, or the mathematical expressions of a design.

Where there is ambiguity, there is personal interpretation.

CORE VALUE #3—PREVENT TRANSLATION ERRORS

In practice, this core value is often more of an "awareness" than anything else. We can identify where translations are occurring (at the very least between each step in the core chain) and understand that translation errors are likely happening. We can then apply various tools (for example, drilldowns, linguistic validation techniques) to root out these errors, increasing the accuracy of the translation.

As we listen to the needs of patients and caregivers, we begin to translate these needs into an addressable need, and the game of telephone begins. As we move farther down the chain, more translation events occur, and we move farther and farther away from the patient we are trying to help. Through understanding and minimizing translation errors, we can help ensure that what the patient or caregiver intended does not get lost along the way.

22

CORE VALUE #4—BUILD LOOKING LEFT, CHECK LOOKING RIGHT

During my tenure at a large medical device company, a consultant was brought in to teach a course on process capability theory for medical devices. She started with a story about a legendary manufacturing company in Japan whose main product was a medical device in the U.S. market, regulated by the FDA. During an audit, the FDA looked into their manufacturing inspection records and found that they were consistently out of tolerance on one of their part's dimensions. The dimension had a tolerance that was inhumanly tight. When pressed by the FDA, the company said that the tolerance was "aspirational," and it was not expected to be met by manufacturing on a consistent basis. The FDA gave them two options: either ensure that their manufacturing consistently meets that impossible tolerance, or open up the tolerance by repeating their design validation testing. The company replied that they would do neither as the changes would be overly burdensome. Instead, they sent a calculation showing that they could be outside of that tolerance by an order of magnitude more than they were and it likely wouldn't affect the patient. As the story goes, the FDA replied, "We didn't create that requirement, you did," and barred the product from being sold in the U.S.

Now I have no doubt that this story had been simplified and embellished over the years, but the point was well made: when creating the product specification, the company was looking in the wrong direction.

THE CORE CHAIN

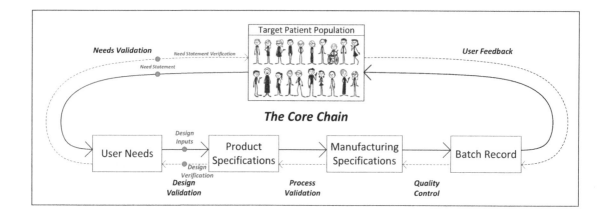

CORE VALUE #4—BUILD LOOKING LEFT, CHECK LOOKING RIGHT

The core chain works left to right, with each major element fulfilling the requirements of the element to its left. Where companies get into trouble is when they create an element based on the element to its right. When I ask an engineer how they set a tolerance on a product specification, I tend to brace myself. If they say, "Well we found . . ." and then go into a description of the product performance, I feel better. If they say, "I called the manufacturer and that's what they feel comfortable with," I cringe.

When you are describing what this product needs to accomplish (user needs), what the product is (product specification), or are building any other part of the chain, you should be looking left, back toward the patient. This is who the product is meant for, after all. The patient (or their caregivers) is making the ultimate determination whether the product is correct, not you, and not manufacturing.

Let's say you have an injection-molded component in your device that you set at 1 inch long. The problem is, exactly 1 inch doesn't exist in nature, so you have to put a tolerance around that 1 inch. It is a classic engineering mistake to call the injection molding supplier and ask what tolerance their manufacturing process can fulfill, say five-thousandths of an inch, and use that as the tolerance. Instead ask, what does the patient actually need? If you end up with a product that is 1.006", is that a problem for the patient? Do you know that 1.004" is not a problem for the patient? What if the patient isn't affected until the length is 1.250", but you just set the requirement 50 times more stringent than it needed to be?

When working on any portion of the chain, the focus needs to be on the previous link in the chain, working to make allowable ranges as large as possible. This principle applies throughout the chain. While creating manufacturing specifications, you develop the largest possible recipe range that still results in product that meets the product specifications, *then* you confirm that the operators and equipment can consistently hit those ranges in actual manufacturing.

When creating user needs, Jon Schell, a fellow medical device consultant and friend of mine, has a mantra: "We focus on outcomes, not solutions." The outcomes look at how the patient will be affected, and the solution is how the product works. Focusing user needs on outcomes makes them agnostic to the design, which allows engineers the freedom to use whatever the most appropriate technology would be to accomplish the requirement.

When using the core value of "build looking left, check looking right," companies align their perspective to the patient and let that flow through their efforts. They give themselves freedom where they can, and only consider their ability to fulfill their requirements *after* they create them.

23

CORE VALUE #5–DELINEATE, DELINEATE, DELINEATE

By September of 1944, the Allies had been consistently advancing into Nazi-occupied Europe, with the Nazi forces retreating to the safety of the Siegfried Line, a hardened defensive line spanning almost 400 miles along Germany's western border. The Allies were starting to beat back the Nazis, but the Nazis' strong defensive line ensured a protracted conflict, with many casualties on both sides.

Operation Market Garden was intended to change that. It was an ambitious operation for a two-day sprint through a main artery of Nazi-occupied Holland and into Germany itself, breaking the Siegfried Line and creating a bridgehead over the Rhine. It would create an inroad into the heart of Germany, hastening the end of the war.

Operation Market Garden consisted of two main elements: "Market"—the largest airborne assault in history, dropping 35,000 men behind enemy lines, carpeting the route, all with the intention of taking the bridges along Highway 69 in Holland, followed by "Garden"—a sprint of the British Army's 30 Corps along the highway, securing the bridges and delivering a knockout punch in Arnhem, 64 miles from the starting point. Timing was critical in the operation. The 30 Corps had to quickly secure ground captured by the airborne assaults, and they needed to be in Arnhem within 48 hours from embarkation, before the enemy could mobilize a response.

Initially, the plan seemed simple enough to be feasible. The 30 Corps had enough resources and support to overpower the Nazi resistance along the route, including the ability to replace almost any bridge that might be destroyed during the conflict. They knew that to be successful, however, a strict timetable had to be followed.

After the massive and surprisingly successful airdrop, 30 Corps embarked on the operation. Right away they found that open ground was too soft for their heavy tank operations, which restricted them to the small, single-lane road. Unfortunately, this made them easy

THE CORE CHAIN

targets when the Nazi infantry and anti-tank guns ambushed them, just 30 minutes into the operation. The single-lane road was also creating frequent traffic jams, as any destroyed, broken, or even temporarily stalled vehicle would stop every vehicle behind it. By the end of day one, they had only advanced 7 miles, 6 miles behind schedule.

It didn't get much better from there. Over the next five days, they were plagued by constant resistance and logistical errors as they marched toward Arnhem, their final objective. On day six (four days behind schedule), the 30 Corps arrived in Arnhem, but found the British 1st Airborne trapped on the other side of the river, being annihilated by Nazi Panzer tanks and infantry. Over the next three days, intense battles, rescue attempts, and positioning and repositioning ensued, all with heavy losses. By day nine, all attempts to take the final bridge had been halted. The force then initiated a retreat and rescue of the British 1st Airborne, retreating along Highway 69, which they started calling "Hell's Highway." In the end, they lost 15,000–17,000 men. The British 1st Airborne, which had been dropped into Arnhem with 10,000 men, retreated with just 2,000 troops left.

Operation Market Garden was, at first glance, clever, bold, and ambitious. They knew that the objectives would be difficult, so they did their best to plan accordingly. The lofty objectives dictated every step of their planning process. They had a careful strategy, a strict timetable, and an intense mobilization of resources. Every conversation they had, every airplane they requisitioned, every soldier they briefed, had those objectives in mind. The problem was that those objectives were just too ambitious, resulting in a margin of error that was too slim. During planning, one of the operational leaders, Polish General Sosabowski, expressed serious concerns about the feasibility of the mission. At one point he famously said, "I think your plan is a bridge too far." He was right.

When considering what your product should be, many ideas and possibilities bubble to the surface. Discussions with different groups lead to different requirements and desires, from the patient, caregiver, and within your company. From these discussions, a lengthy shopping list of possibilities for your product emerges, with dozens of potential design targets. Sales would like the unit to cost this much, a particular doctor would like to see a feature similar to a competitor's device, and marketing would like to put branding on every square inch of the product and packaging. Are all these requirements created equal? Of course not.

CORE VALUE #5—DELINEATE, DELINEATE, DELINEATE

Every User Need on the core chain will be an all-important objective for your product and your company. They are what you must accomplish with your product, and are the primary measure of success or failure that you are proposing to the FDA, prior to product release in the market. Every objective you add requires further complexity, resources, constraints, validations, and quality control. This is not just true for the initial launch, but throughout the life of your product. Not meeting the user needs is not an option; you don't want to be in a situation where you're explaining to the FDA why you changed a User Need to reflect your product's performance, rather than the other way around. They will (and should) say that this demonstrates that you don't understand your user's requirements.

So, what is on the user needs list and what isn't? Well, first let me say that the user needs list is not the only list that we can make. When the engineers are designing the product, there's a lot to communicate about what the product needs to accomplish and what you want it to accomplish. I like to use a few lists to capture these product goal differences. While these are lists that I have used in the past, this is more of a build-your-own-adventure situation.

DELINEATED PRODUCT GOALS LISTS

User Needs List—These are the core chain requirements, and they *must* be accomplished by the product. These needs are often tied to safety and efficacy, as well as regulatory and statutory requirements. Not only do you need to absolutely achieve these, you must provide *objective evidence* that they have been accomplished. Any change to the product specifications, manufacturing specifications, or batch record must be evaluated against these requirements.

Business Requirements List—Business requirements lists are requirements that are of critical importance to the company but are not tied to safety, basic efficacy, or statutes. The unit cost where the product is no longer financially feasible is an example. This has nothing to do with the safety or basic efficacy of the device, but the business might not go to market if a certain price target is not met. While these are requirements, they are handled differently than user needs. No objective evidence is required (although the business may want it), and if a business requirement is not met, it is entirely up to the business on how to handle it, including disregarding the requirement.

Business Wants List—Often coupled with a "scale of strength," the business wants list is the laundry list of product goals. You can go to market without the business wants, but the product may not be as successful. The criticality of an item on this list is often hotly debated, with marketing and engineering on opposite sides of the room like a middle school dance. Once again, these are not required to be shared with anybody outside of the company, and you can define how you determine if the "wants" have been met.

A smart capability of delineation is that the same feature can appear across multiple lists. For instance, a User Need may be that an implantable battery lasts a minimum of three years, a business requirement that it lasts five years, and a business want that it lasts seven years. Each one of these has a different proof burden and different level of flexibility in the requirement.

THE CORE CHAIN

These cascading objectives on the same feature can help frame the development efforts, and the quality response to any issues related to that feature.

That's what it comes down to: correctly setting and framing objectives, not only in the initial development of the product, but throughout its entire life cycle. The amount of resources you need and your margin for error will depend largely on what requirements you set for yourself. It may be the difference between historic failure to capture nine bridges, or a successful operation to capture eight.

24

CORE VALUE #6–THE SMALLER THE LINKS, THE TOUGHER THE CHAIN

Anyone who has written a detailed process software validation will tell you they're a hoot. Tons of fun. Boatloads of fun. Very similar to a night on the town, so long as that town is a solitary shack with nothing in it but tedium, and the night lasts dozens or hundreds of hours.

It used to be simpler. Companies used to do "black-box" testing, which was basically the idea that if the manufacturing process worked, then the software running the equipment must be working correctly. Those were easy as pie to write. You didn't need to know much about the software, or the process really. You were in effect saying, "We don't know what's happening behind the scenes and we don't care." Those were simpler times. And worse times. Much, much worse times.

Black-box testing lumps the software into one complete, unknowable entity, that is, "the black box." No attempt is made to understand what the software is doing at the fundamental level, or the likelihood that something deep within the software could go wrong. The black box was tested through the general functionality of the equipment, then the black box was locked down, preventing any changes.

With the black-box approach, one of two situations would eventually occur. First, under some set of circumstances, the output of the black box would be wrong. The inputs would seem correct, and the logic of the black box should hold true, but the system did something unexpected. As you don't understand the intricacies of the black box, this ends up being deeply troubling. In an industry where consistency is king, the last thing you want is an unpredictable result in your manufacturing process. The second situation was more common. When something on the equipment changed, either planned or unplanned, the entire software validation was called into question. With no visibility into the inner workings of the software, if it had to be altered, even in a minor way, you could not justify that the change only impacted a portion of the processing.

THE CORE CHAIN

As changes to these pieces of equipment always occur (components would become obsolete, modifications to the equipment would be required, and so on), you were back to square one after any sort of modification. An entire revalidation was required, and you still didn't have a good road map for how the software operated.

At first glance, the black-box style of testing is attractive. It is less expensive, faster, and if you are testing the system as a whole, why would you care to understand what is going on at a fundamental level? This has bitten us hard enough and for long enough as an industry that black-box testing is now considered amateur hour. Current best practice is to fully describe the software design and test the individual activities performed by the software. This type of software validation is sometimes called *deep testing* or "white-box" testing. This approach requires more effort spent understanding and characterizing the software, but it pays dividends in the end, putting you in a much more defendable position while allowing for future flexibility with less pain.

The black-box testing versus white-box testing idea extends far past software validation and is, in fact, the exact same case with every link of the core chain.

CORE VALUE #6 — THE SMALLER THE LINKS, THE TOUGHER THE CHAIN

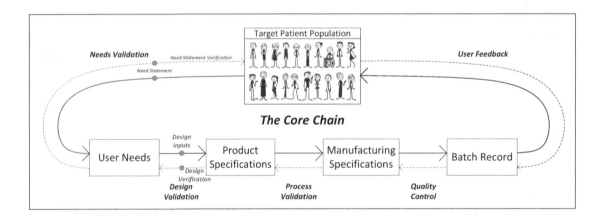

When looking at two sections of the core chain that need to be linked, there are a few ways to approach it. Let's take user needs and product specifications, for instance. We have a list of product specifications that we need to validate against the user needs. Let's say we do this only through a clinical trial. We have the entirety of a product that we test against all user needs at once. If we are successful, then all product specifications, as one, are validated against all user needs, as one. We have one single link between everything in this part of the chain. We have no visibility into how one product specification affects one User Need, or any important combinations or interactions between these needs and specifications. If we set a product specification length to 1.000" +/– 0.010" and we want to change to 1.000" +/– 0.015", then the entirety of the clinical trial results, and the proof that you have fulfilled the user needs, could be called into question. This style leads us to the "one bad, all bad" result. If the only resolution you have is that the totality of your product fulfills all user needs, your chain will easily break.

The more you individually trace elements of the chain to each other, the stronger the overall core chain becomes. Below is our core chain example from Part II.

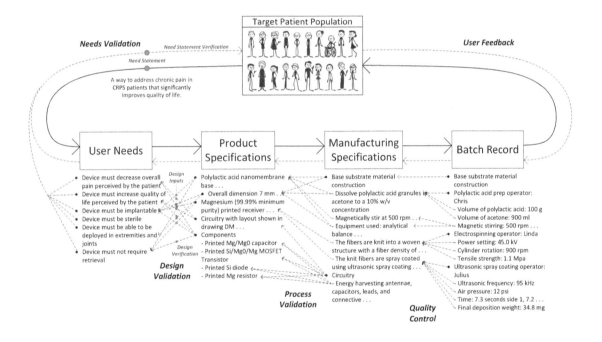

In this example, individual elements of each step in the chain were traced to individual elements in the prior link in the chain. So, if a manufacturing engineer thinks an upgrade from a 45 kV to a 55 kV electrospinner would improve the manufacturing, you can trace which product specifications this change might affect and then revalidate against only those pieces.

The more granular these links are, the healthier your overall chain. This granularity requires (and demonstrates) a deep understanding of your product and its interaction with the patient

THE CORE CHAIN

and manufacturing. Detailed links allow us to understand and thus justify any future changes to any sections of the chain, and to better diagnose issues seen in manufacturing or the field.

The problem is that there is a limit to how much time and resources you can employ to understand these individual relationships. I get it. There are always deadlines, external and internal pressures, and a desire to stay as lean as possible. This is one of those areas, however, where effort spent on this stage pays dividends throughout the life of your product.

Companies that take this value seriously consider the balance that they need to strike. Performing intense, deep characterizations on low-impact sections of the chain probably doesn't make sense when you're in a race to the market. Luckily, much of the chain will not require this level of deep analysis. Companies often use mapping tools such as trace matrices or failure mode and effects analysis to show how these elements link (or don't link) to other elements on the chain. Rationale is documented as to why links don't exist between items (for example, printing a logo on the device isn't tied to the overall length), and where links do exist, the relationship is characterized with a level of rigor commensurate with the risk.

These trace matrices become the road map for future quality activities and act as a storehouse for additional understanding of these links over time. Doing this ahead of time, rather than justifying inaction after an issue or change is seen, allows the opportunity to truly use it as a road map for the management of the product over time. The merit and granularity of this road map can directly affect the future quality of life for your operations personnel, development engineers, and quality team members tied to this product.

25

CORE VALUE #7 – ROUGH IT IN AND REFINE

In my last semester of my undergrad, I needed a general class to fill out my credit requirements, and I decided to pick something outside my wheelhouse, Drawing 101. The instructor was a mid-20s guy with a very relaxed and pleasant demeanor. The class would travel to various interesting locations around the area, and we would sit and draw what we saw, using pencils and sometimes charcoals. While sitting with my drawing pad in the woods outside the architectural wonder named Fallingwater, the instructor sat next to me and described what he saw as I drew. He was describing the shapes, the lines, the relationship of the light and shadow, and the combination of motion and stillness. Due to my complete lack of artistic ability, I couldn't properly capture any of that, but it was amazing to hear what his eye was catching.

The thing that struck me most about that class was the use of erasers. Our instructor, along with his students, would have a pencil in one hand, and a big tri-tip eraser in the other. When he would draw a line, he would look at it, and then erase it. He would then draw it again, slightly different, and erase it again. This would happen again and again until his line was just right. Then he'd move on to the next line. He would adjust that one until it was just right, and might go back to the previous line if the relation between the lines was off. He was constantly drawing and erasing, drawing and erasing. He probably erased three lines for every one he kept. I had assumed that a skilled artist, as he was, would be able to draw exactly what he saw on the first try, but that's not at all how it worked. Slowly, I realized that the ability of his hand, while far superior to mine, was not what made him a skilled artist; it was his eye.

THE CORE CHAIN

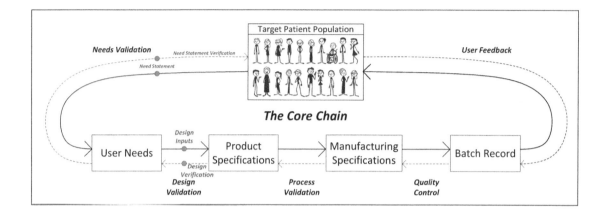

CORE VALUE #7–ROUGH IT IN AND REFINE

When going through the stage-gate product development process, you typically stay in one stage until it is fully completed. As a matter of fact, a primary purpose of a gate is to prevent you from moving to the next stage until the previous stage is fully done. This typically results in a serial waterfall-style approach, where one element of the core chain is fully created before moving on down the chain, often handing the baton to a different group to continue the progression. This is a clean project management tool, as you can check items off the list and consider them done, showing your team's progress down the path. The sequential waterfall approach assumes that everything you have built is correct and will fit together nicely in the end. I have never been a part of a project where that assumption is anywhere near correct, and the approach is, well, let's say . . . suboptimal.

While our product development process is laid out sequentially, even *chronologically*, the core chain is different. The core chain follows a *logical* framework that is not meant to be completed one step at a time. The picture is too complex and relational to be built this way.

25—CORE VALUE #7—ROUGH IT IN AND REFINE

Fallingwater by Anna Hryshchenko

Building one element of the chain before moving to the next one would be as if my drawing instructor started on the far left edge of the drawing and only worked within the first inch. He would need to stop drawing any lines that went beyond that inch, and he would need to complete all the structures, details, and shadows within that inch before he could move beyond it. What's worse is that the next inch would need to be created by someone else, who would then hand it off to someone else when their portion was done. This would be a mess. The relationships within the drawing would be inconsistent, the lines would be distorted, and the overall impression would be muddled. It might make for an interesting drawing, but it would not be the best way to capture what you're trying to capture.

Of course, the artist does not do this. They start by sketching the overall structure across the entire canvas, then adjusting it over and over again until it looks right. Only then do they add

THE CORE CHAIN

the detail to the structure and adjust it again if something isn't quite right. The artist builds and rebuilds, adding more detail when they're sure what they see is correct. They see the picture in its entirety, making sure lights and darks, perspective and lines, motion and stillness, all work together seamlessly, creating the impression they want.

The core chain must be built the same way. There are many complex relationships between and within different elements of the chain. As we progress along the product development process, we must build and rebuild the entire core chain, roughing in the broad strokes, fixing what isn't right, then progressing only when our critical eye says it's time. This is one of the primary differences between the traditional waterfall product development process and the core chain method of product development. We will explore this in more detail in Part IV.

26

CORE VALUE #8 – CHECK AND PROTECT

If you look at the skyline of New York City, you'll see a towering array of globally unique, instantly recognizable skyscrapers that look like they are rising out of the water. The Empire State Building, the Freedom Tower, the Chrysler Building, the Woolworth Building, and Rockefeller Center are just a few in this chorus of giants. When admiring the Empire State Building in Manhattan, you may see a skyscraper that seems different than the others; a 59-story aluminum-and-glass building, with a stark 45-degree angled top. This is the Citigroup Center, built by Hugh Stubbins and William LeMessurier, which, at one point, was in danger of destroying most of midtown Manhattan.

When you approach the Citigroup Center, it isn't the angled top that surprises you, it's the bottom. When Citigroup bought the block, one church, St. Peter's Lutheran Church, was not interested in moving. St. Peter's was in rough shape, however, and they eventually agreed that if Citigroup would build them a new church in the same location, they could have the sky above them. Citigroup agreed, and one of the most interesting engineering challenges was created. LeMessurier was up to the task, designing the entire 59-story building to sit on nine-story-tall stilts, creating a beautiful yet surreal open-air feeling at the large base of the building. The building was completed in late 1977 and was considered an engineering marvel, with the title of seventh tallest building in the world.

In June 1978, LeMessurier received a call from an engineering student, Diane Hartley, who was writing a thesis on the Citigroup Center building. She expressed concern that because of where the stilts were placed, quartering winds (winds that strike the building at its corners) could topple the building. The concern was instantly dismissed by LeMessurier, and he moved on to other matters. A few days later, however, he decided to reconsider the student's concerns and re-perform the stress calculations on the building. As he started to dig into the problem, he realized that the engineering student might actually be right. LeMessurier had originally calculated that a 1-in-500-year storm was required to topple the building. Based on his new calculations, it would only take a 1-in-55-year storm to destroy the building. The worse news was that if the tuned mass damper (a 400-ton floating concrete ball on the roof) wasn't working properly, or if the building lost power (which often happens in a major storm), it would only take a 1-in-16-year storm to cause unimaginable catastrophe. This reframed the fall of the Citigroup tower from a near-impossibility to a matter of time. Notifying anyone of the issue was exclusively up to LeMessurier, as he was the only one who had full insight into the problem.

LeMessurier rushed to meet with Citigroup executives with full transparency of his error. The group quickly appreciated the gravity of the situation. In an attempt not to alarm the people occupying the building, they created a plan to fix the structure quietly, with welders fixing the joints through the night after the staff had left the building.

In August, as they were halfway completed with the fix, a hurricane formed off Bermuda and was headed straight for Manhattan with winds fast enough to easily topple the building. As they had already been notified by Citigroup of the issue, the NYPD and Office of Emergency Management had an evacuation plan prepared that spanned a 10-block radius. Twenty-five hundred Red Cross volunteers were placed on standby, along with three different weather services, all in secret. In a stroke of luck, the hurricane performed a last-minute turn back out to sea before it reached New York, and the repairs were completed without further incident.

LeMessurier's actions are now considered a case study in professional ethics. The fact that he reconsidered his calculations when challenged (especially by someone at the beginning of their career, when he was at the pinnacle of his), and how he reacted after finding the serious flaws in construction, turned a potential disaster into an example of ethical behavior. While LeMessurier was not perfect, by the Red Cross's estimation, his actions and the actions of Diane Hartley may have prevented a death toll of an astounding 200,000 people, as well as the destruction or serious damage of 156 city blocks of midtown Manhattan.

Once we have built something we are proud of, it can be hard to look at it objectively. In the product development process, we will spend considerable time and effort getting to the enviable point where we can sell the first product. To then start looking at your creation with a critical eye requires a force of will, a healthy dose of self-confidence, and an understanding that it is your obligation to do so. LeMessurier was able to do it, albeit with some reluctance, and he helped avert a monumental tragedy.

If done correctly, the ability to sell that first product is the signal that the core chain has officially been built. Our product development project is complete, and our product goes into a new stage, often called "post-market." If we have built a product that patients need and the market wants, we should be in this stage for a long time, ideally 10 or 15 years plus. The team managing the product often changes as we switch to post-market, and the systems

26–CORE VALUE #8–CHECK AND PROTECT

and metrics we use to govern the product will shift too, as we are no longer building our core chain, we are protecting it.

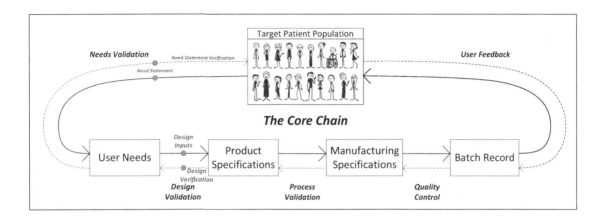

To get to the point of saleable product, we have established and documented the core chain, and we have proposed the risk/benefit equation to the FDA and the marketplace. If we were lucky enough to get that accepted, we need to ensure that the core chain and risk/benefit equation does not change from what we promised. To do this we must have tight enough control to manage and characterize any changes prior to the product reaching the patient. Easy enough, right?

Anyone who has manufactured medical devices knows that this is a difficult proposition. While we have more direct control on our manufacturing floor, issues that can influence our product go far beyond our walls, and change is certain. The core chain that we spent so much time building and characterizing will break. It is inevitable. The degree to which it will break is a question, but, unfortunately, it will break. How it will end up broken, luckily, is well understood.

The reason why your chain is broken usually falls into one of three categories:

1. The core chain was never built correctly in the first place.
2. Something changed (unintentionally).
3. Something changed (intentionally) and it wasn't rebuilt correctly.

As I said, links in our chain will break. One of our challenges is that in practice, this chain that we are relying on is invisible. We can map out the logic, we can map out the correlation between sections of the chain, but we cannot always see if one of the all-important links has been broken.

THE CORE CHAIN

CORE VALUE #8—CHECK AND PROTECT

The integrity of our core chain is paramount. It is the only way we can have confidence that the product we are sending to the patient meets our safety and efficacy requirements. The protection of our chain is our responsibly, as is our vigilance in examining it for breaks. There are many tools we can employ to check and protect our chain, most of which can come together as a quality management system. The two primary tool types we use are proactive tools and reactive tools.

Protect the Chain (Proactive)

To protect the core chain is to prevent breaks in the chain before they occur. The core chain is delicate, and to protect it from breaks requires a complete set of systems. Document control, change control, calibration, training, facilities and equipment, preventive actions, and many more systems have one primary purpose: to protect the core chain. Once the chain has been established, it represents the correct version of everything we have developed and tested. If anything changes, then the integrity of the chain is in question. Most of these systems ensure that you are operating within the bounds of the core chain as intended.

Check the Chain (Reactive)

Checking whether the chain is broken is a reactive stance, but it is necessary to understand whether you have lurking issues. Quality control is the first line of defense on this: performing inspections and chasing down nonconformances to see if anything is out of control. Corrective and preventive actions (CAPAs) are the second line of defense, either showing that the chain is broken, or that a process issue may lead to a break in the chain. The third and last line of defense is user feedback. User feedback is a combination of adverse events, complaints, general feedback, or any response from the field. This feedback from the users is very late in the process and can have major impacts on your company, so ensuring that checks are in place prior to the product reaching the users is crucial. That said, user feedback is the most complete, unfiltered, and unbiased feedback on the product and your core chain, so it must be treated with reverence.

If you are fortunate, your product will be highly marketable, and you will be in a check-and-protect mentality for a very long time. You will find that every ounce of effort and care that was placed into building the chain will be rewarded here, and all previous sins will resurface. As we live in our post-market world, it is important that we deepen our understanding of our product's core chain, challenge assumptions we made along the way, and remain vigilant to its integrity. By doing this, our product, and the positive impact it will have on the patient, will strengthen over time.

27

PART III: RECAP

Just to beat a dead horse, here is a quick cheat sheet on the core values that expand on the core chain.

1. **Build the Right Chain**—Just because you have a great idea doesn't mean it's useful, and just because an idea is useful doesn't mean it fits within the competitive landscape. Understanding where your product fits in the landscape of choices to the patient and caregivers, their perception of usefulness, and the competitive dynamics you can expect, all help ensure you are building a product that people will actually want.

2. **Build from Left to Right**—The farther left on the chain you are, the more critical it is to be correct. All activities downstream (to the right) are using the items on their left as their target. If the targets are incorrect, the effort to fulfill those targets is wasted.

3. **Prevent Translation Errors**—The translation of a need statement into user needs, user needs into product specifications, and so on farther on down the chain, results in a "telephone game" of sorts, where miscommunications and biases have a large impact on results. These natural translation errors need to be battled all along the chain.

4. **Build Looking Left, Check Looking Right**—Every step in the chain needs to be built looking toward the patient and subsequently checked downstream on the chain. This is often difficult for engineers, thus leading to a classic mistake. It is important to give your company freedom where you can, and to build requirements without worrying about your ability to fulfill them.

5. **Delineate, Delineate, Delineate**—Items that are on the core chain are items that affect patient safety and efficacy. If it does not affect safety or efficacy, then it does not belong on the chain. Items on the chain need to be scrutinized to a high degree

of assurance, and items that do not need that level need to be handled separately, regardless of business importance.

6. **The Smaller the Links, the Tougher the Chain**—Companies run into trouble when items are tied out at a macro level (all these engineering specifications together fulfill all the user requirements together), rather than a micro level (this one product specification has this effect on one User Need). When this is done at a macro level, if one specification is bad, they are all bad. The more granular the connections are, the tougher they become.

7. **Rough It in and Refine**—While the typical stage-gate product development process is chronological, the core chain is not. The architecture of the core chain is to be roughed-in overall, then levels of sophistication increased, left to right. The overall picture should be seen from the beginning and adjusted and refined as you progress through the product development process.

8. **Check and Protect**—A completely built core chain is a beautiful thing. The core chain is your responsibility, and once it is complete, you should be diligently checking for breaks and vigilant in its protection.

Individually, these core values help guide any company building a medical device. When combined, built on the core chain framework, we can refine our approach ever further, and an opportunity for a very different product development pathway presents itself. We will describe this approach in Part IV.

PART IV

THE CORE CHAIN METHOD OF PRODUCT DEVELOPMENT

Part I	Part II	Part III	Part IV
Developing a Medical Device Using Traditional Techniques	The Core Chain Framework	The Core Values	The Core Chain Method of Product Development

Combining everything covered to this point, let's redo our efforts to commercialize a medical device. We'll start from the beginning when you joined the company, and then we'll reconsider the journey of product development using our new approach, the core chain method. Through the process we'll cover how the core chain method is organized, as well as the fundamentally iterative nature of this approach.

28

WELCOME ABOARD

We're so thrilled you decided to join our new start-up! Welcome to your first day at our little company. I know you're as excited as we are to bring this new medical technology to market. We've been anxiously awaiting your start date, and we feel we can officially start our product development activities now that you're on board. I know today is your first day and there is a lot to do, but as you are part of the leadership group, I think you and I should spend the morning together.

Now, during your interviews we gave you an introduction to the idea we're hoping to commercialize. I'm guessing it's a big reason you decided to join us. But a great idea isn't enough. Have you heard the expression, "Great ideas are common, great execution is rare"? Well, nowhere is that statement more true than in the world of medical technology development. If we want a real shot at our technology becoming a reality, we need to dramatically enhance our development approach. I'm not just talking about a few tweaks here and there; I'm talking about a fundamentally different methodology to medical device development. Intrigued? Good, why don't you get comfy, maybe grab a double-latte, and I'll tell you all about it.

29

INTRODUCTION TO THE CORE CHAIN METHOD

Now, you and I cut our teeth on the traditional medical device development process, the waterfall method.

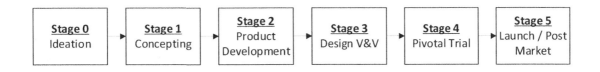

The Waterfall Method of Medtech Development

We've spent countless hours executing this approach. Writing project plans, designing our products, carefully verifying and validating our devices. Through the years we've understood the pros and cons of this approach, and while we complained about it, we didn't really know of another option. Well, I'd like to introduce you to the *core chain method* of product development.

THE CORE CHAIN

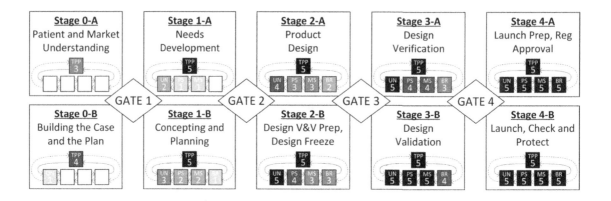

The Core Chain Method of Medtech Development

The core chain method is a new approach to medtech development, and while it keeps many of the same individual activities as the waterfall method, it is rebuilt around three key focus areas: *visualization*, *alignment*, and *iteration*. I'll elaborate on each of these in turn, then walk you through how they all fit together. Once we're done with that, you and I will talk about each individual stage and substage in the process, and list the activities we'll need to perform to progress our core chain.

With that, I'd first like you to think about **visualization**.

I wonder if you've noticed, as I have, how difficult it is to visualize what we're doing while we're developing a medical device. We can lay out all the tasks in front of us, track how complete they are, even see how the physical product is evolving. But few seem to visualize the actual progress toward our ultimate goal, or even what our ultimate goal is. This causes all sorts of problems during execution: working on activities at the wrong time, an overestimation of where we are versus where we need to be, and a general misunderstanding about what "complete" looks like. To improve our execution, we must first understand and visualize our goal, and be able to map our progress toward it.

To do this, we're going to leverage a concept that I hope you're familiar with at this point, our old friend the core chain.

29 – INTRODUCTION TO THE CORE CHAIN METHOD

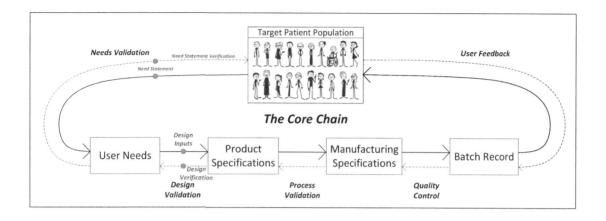

The core chain will be our north star for the project, a finish line and guide to our activities. We will use it throughout our development process to anchor and organize our thinking. But I want you to understand that the core chain is not a binary structure, something that either exists or doesn't. I want you to see it as an intricate creation, with large, interrelated elements and small details that work in combination. Let's look again at the drawing of Fallingwater by Anna Hryshchenko from our *rough it in and refine* core value:

Fallingwater by Anna Hryshchenko

THE CORE CHAIN

To create this drawing, the artist (Anna) started by building the basic shapes, sketched out with loose lines that gave form to the idea. These shapes and lines were refined until the overall structure looked correct. She then drew hard edges to create the transitions between objects, providing depth and structure to the drawing. From there, Anna built on this structure with detail, adding soft lines, shading, and hatching until the final form took shape. She then stepped back with a critical eye and refined, erased, and fixed areas until she created the impression she was striving to achieve. Anna did not just work directly on the final drawing, she *progressed* through the stages needed to bring it to its end state. This progression of stages from a blank page to a fully detailed work of art is similar to the progression that your core chain must undergo.

While the progression of a drawing is easy to see, the core chain is more opaque. To map our progression, we must understand and define how the core chain should evolve through the development process. The core chain method achieves this through *sophistication levels*. These sophistication levels describe the evolution of each element of the core chain from a blank sheet of paper to a fully detailed end state. The basic structure of the core chain method uses sophistication levels of 1 through 5 for each element, which we can use to visualize our progress toward our destination (sophistication level 5). While each core chain is unique, we can use the following sophistication levels as a starting point for each of our core chain elements:

CORE CHAIN SOPHISTICATION LEVELS

	Target Patient Population (TPP) Sophistication Levels
1	General target patient population identified and described.
2	Typical patient profile created. First draft of need statement created.
3	Target patient population well defined with clear borders. Possible subpopulations and patient journey identified. Need statement challenged and further crafted.
4	Target patient population characterized with defined subpopulation segments with profiles. Target patient population definition is validated.
5	Final target patient population is thoroughly characterized, described, and validated. Need statement is finalized and validated.
	User Needs (UN) Sophistication Levels
1	Primary efficacy and indication-specific safety user needs defined but potentially non-specific. Non–core chain requirements delineated off user needs list.
2	More complete efficacy and safety user needs list is created. Initial testing/characterization of user needs.

29—INTRODUCTION TO THE CORE CHAIN METHOD

3	Regulatory and statutory requirements added. Efficacy and safety needs more deeply characterized with light verification by SMEs or individuals within the TPP. Initial design inputs are drafted.
4	All user needs fully defined and specific. Borders of individual user needs challenged. user needs have been verified against the TPP and associated caregivers. Design inputs are fully detailed.
5	All user needs are fully defined and have been objectively validated against the TPP and associated caregivers. Design inputs are verified against the user needs.

Product Specifications (PS) Sophistication Levels	
1	Popsicle sticks and pipe cleaners. Looks like you made it in kindergarten. Possibly some digital concepts.
2	More sophisticated concepts. Basic prototyping methods (3-D printing, and so on) to create physical representations of the product. The specific concepts are tested for specific types of product feedback.
3	Physical representations (although using some different processes than the final production processes) have been created. Product specifications are complete. Product has not been fully characterized.
4	Product has been built and tested using scaled-down versions of the final production processes. Product specifications are fully detailed but not fully challenged.
5	Final product specifications, including software designs, are complete, verified, validated, and frozen. Actual production-equivalent units of the product exist.

Manufacturing Specifications (MS) Sophistication Levels	
1	Basic process flow created showing all the major steps involved in making the product.
2	All process steps captured and possible equipment and process materials defined. Major process risks identified.
3	Detailed process flow created with all materials ins, outs, process variables identified. Test methods identified. Preliminary equipment defined. Pilot line created.
4	All processes and final equipment defined. Final equipment installed and IQ'd/OQ'd. All processes and equipment fully characterized. Test methods fully defined and described.
5	All individual processes, equipment, and test methods fully developed, characterized, and validated.

Batch Record (BR) Sophistication Levels	
1	Basic process flow walked through by operations/experts.
2	Basic assembly captured and attempted—not necessarily on production equipment.

THE CORE CHAIN

3	Initial batch record/work instruction set created and tested by operators on early production equipment or surrogates.
4	Full production batch records/work instructions created, executed, and challenged by operators and engineers on production equipment.
5	Full production batch records/work instructions created, executed, and challenged by operators and engineers on production equipment. Entire production PPQ'd.

With our sophistication levels defined, we can not only visualize where we are in our core chain development, but also every major progression we must undergo on the way to our final, fully detailed piece of art. At any time, we should be able to describe the status of our core chain visually as shown below.

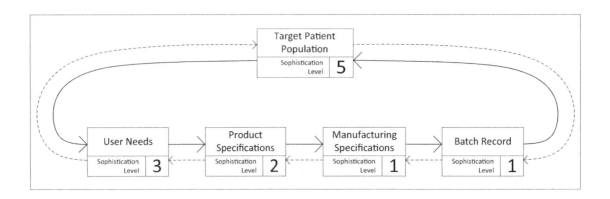

Example Core Chain with Sophistication Levels

We can see that the Target Patient Population element is complete, the user needs element has a sophistication level of 3, and so on. Furthermore, we can now describe the activities we need to perform to increase the sophistication level for any one of these core chain elements. For instance, if I want to increase the sophistication level of the product specifications from a 2 to 3, I know we need to: create a revision A engineering drawing set, design and build complete preproduction quality devices, test preproduction quality devices, and perform the design and use-risk assessments (more on that later). When we complete these activities, our product specification sophistication levels up, and we update our visual on where we are:

29—INTRODUCTION TO THE CORE CHAIN METHOD

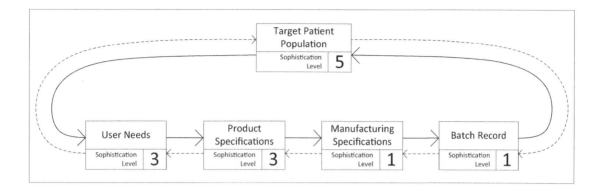

At the end of our product development process, we'll have a 5 for each sophistication level, which signifies that our core chain is complete. We can use this method to organize all our product development activities based on how they help build each one of the elements of our core chain. This visualization and tracking of the progression of the core chain is one of the major tenets of the core chain method of product development.

The next natural question is in what order should we level up the elements of our chain. That brings us to our next key focus area: ***alignment***.

Using our visualization and organization methodology above (building our core chain campus, one brick at a time), our team should be internally aligned to what we are building, how far along we are in building it, and what activities we need to perform to get where we want to be. While our activities are aligned internally, that does not mean we are aligned with the target patient population, or their caregivers. To build that alignment, we're going to rely on one of our core values, *build left to right*.

We can now see the level of sophistication of each of the elements of our core chain, and we know the importance of building left to right, so this can help direct us on the order in which we level up our elements. If our core chain is at the levels of sophistication shown in the chart below, it would make no sense for us to level up our product specifications until our user needs are further along.

THE CORE CHAIN

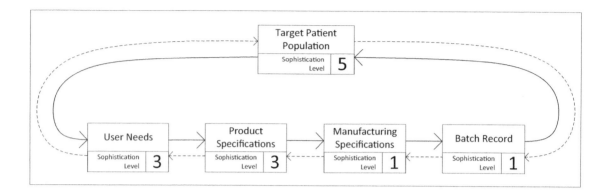

As we know, that would be a recipe for issues, as we would be progressing too far along defining solutions before we define our desired outcomes. This could lead us to lose alignment with the target patient population and their needs. Thus, part of the core chain product development methodology is to *always have an equivalent or higher level of sophistication on the element to the left*, starting with the target patient population. This prevents us from working on aspects of the chain at the wrong time, and helps keep us aligned with the target patient population and their caregivers.

With that in mind, wouldn't it be easiest to build up the target patient population to a level 5, then start on user needs, get them to a level 5, then progress down the line? Nope, quite the opposite, actually. This brings us to our last key area: ***iteration***.

As you recall, the traditional waterfall method generally relies on a sequential approach to product development. One team is primary on the first piece of the chain, they complete it, then you move on to the next major activity, with another team taking over. Marketing owns the user needs, R&D owns the product specifications, operations owns the manufacturing specifications, and so on. As you know from the *rough it in and refine* core value, this is equivalent to drawing a picture left to right, fully detailing one inch at a time. Perspectives are siloed, translations are wrong, and the overall picture can get skewed along the way. To battle this, we're going to integrate our *rough it in and refine* core value into our product development process. We do this by making our core chain development *iterative* overall.

29—INTRODUCTION TO THE CORE CHAIN METHOD

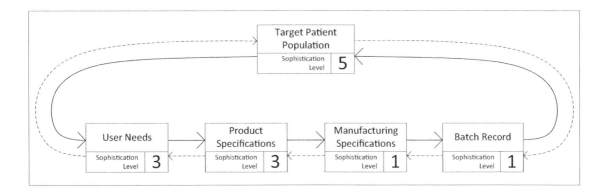

By taking this approach, we will still increase sophistication left to right, but we will continue down the chain over and over again, flowing our new understanding down the chain, understanding its impacts downstream, then reconsidering and updating the left elements of the chain, flowing that down the chain, and so on. In other words, we're defining our desired outcomes, trying out a solution, then stress testing that solution again and again with progressing levels of sophistication. This iteration sequence is fundamental to our core chain approach to product development. Looking at the core chain sophistication levels in our previous graphic, I'd think we're getting a little heavy on the left side, and would probably level up the manufacturing specifications before moving back to level up user needs. Keeping the chain in balance (while still somewhat lopsided to the left), helps us gain insight into the implications of our decisions that help us build a better product. So what does this actually look like in practice? Well, here is the basic map for our development of the core chain:

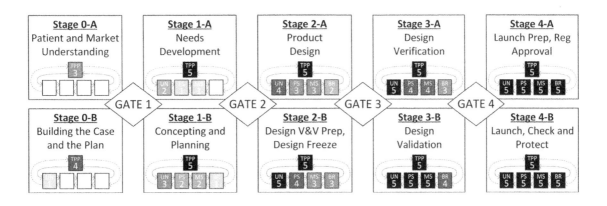

THE CORE CHAIN

Behold, the core chain method of medical device development. You can see I converted the larger picture of the core chain sophistication levels to a bite-sized version, but the idea is the same:

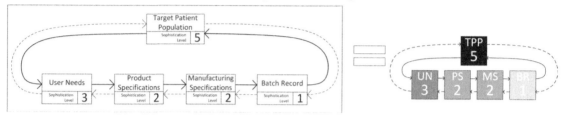

This way we can see the target core chain sophistication level at the end of each substage in our stage-gate process. So, for instance, at the *end* of substage 1-B, our target patient population should be at sophistication level 5, our user needs at level 3, our product specifications at level 2, our manufacturing specifications at level 2, and our batch record at level 1.

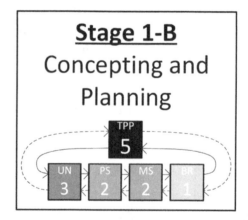

With this, we can see how our core chain should evolve over the life of the program, and we can organize our activities into the substage where they're needed to level up an element of the chain. And we'll do exactly that. As we consider each stage in the process, I'll provide you with an introduction to all the activities needed to level up the elements of our core chain in that stage.

29—INTRODUCTION TO THE CORE CHAIN METHOD

Note to reader—Part of the intent of Part IV is to be a "reference guide" to the activities needed to complete the core chain method of product development. If an understanding of these activities is not yet required for where you are in development, feel free to skim through them and come back at a later date when they are more relevant.

Before we get started on the stages, I want to mention a few best practices that we will be implementing as part of the core chain methodology. First, those insights we talked about gaining from our *rough it in and refine* iterative approach could be easily lost if the project is handed off between departments, or even between team members. We're going to want to take full advantage of these lessons we're learning about our patients, the product, and the manufacturing as we iterate along the chain. So, we're going to have a cross-functional team building the chain together from the very beginning, with as much continuity through the development process as we can muster. This cross-functional team will build all elements of the chain together, which will maximize our knowledge and help us combat various forms of bias (solution bias, past-experience bias, and so on.), that naturally pop up during the process.

Another best practice you'll see along the way includes our periodic *build the right chain* challenges. This is part of our alignment initiative. As we become more and more knowledgeable along the way, we'll step back and make sure we are building a device that the patients and other stakeholders actually want.

The rest we'll figure out as we go. I hope you're as excited as I am to get this idea on the formal path toward becoming a reality. Before we get too far along, however, I'd like to tell you about the stages we'll be traversing along the way.

30

THE CORE CHAIN METHOD–STAGE 0

As we discuss each stage, I'll remind you where we are on the overall journey, as well as how far along the core chain we should be at this point. I've made some handy visuals to help keep us on track, but I know they are new to you, so let's take a minute to explain how the visuals work.

Again, here is our overall core chain method for medical device development:

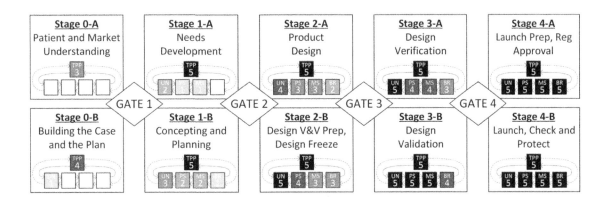

(TPP—Target Patient Population, UN—user needs, PS—Product Specifications, MS—Manufacturing Specifications, BR—Batch Record)

You can see that each stage is broken up into two substages, with the *ending* sophistication level for each element of the core chain shown at the bottom of the box. Each one of these substages has a little different focus, but most are covering multiple parts of the core chain

THE CORE CHAIN

through the activities. Now, at the beginning of a stage, we'll break out the two substages and go through them, one at a time. For instance, here's the substage 0-B detail:

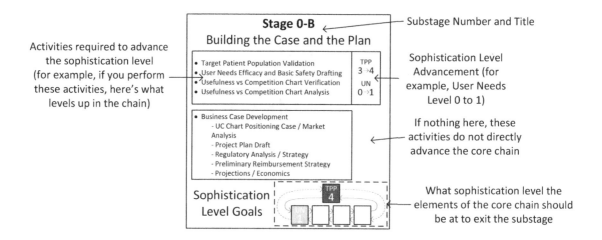

You can see that it's organized by the activities that are required to advance the elements of the core chain. You may notice that each core chain element becomes darker as we advance its sophistication. This allows us to see at a glance how our core chain campus is shaping up. So, in the Stage 0-B example, we know that performing the Target Patient Population Validation, user needs Efficacy and Basic Safety Drafting, Usefulness versus Competition Chart Verification, and Usefulness versus Competition Chart Analysis will level up our Target Patient Population element from a level 3 to a 4, and our user needs from a 0 to a 1. When we're complete with Stage 0-A, we move on to Stage 0-B, and when we're complete with Stage 0-B, we'll go through a gate before moving on to Stage 1-A.

With that primer in mind, we're officially at the starting blocks. Welcome to **Stage 0** of the core chain method.

30—THE CORE CHAIN METHOD—STAGE 0

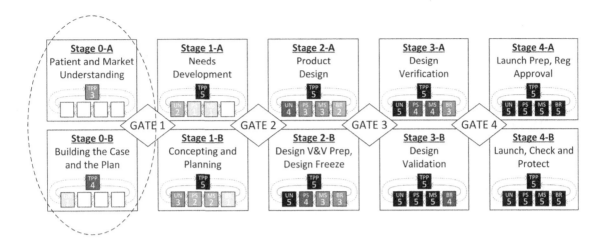

There are two substages that are laid out for Stage 0, each with a little different intent. The first substage is: **Patient and Market Understanding (A)**; and the second substage is: **Building the Case and the Plan (B)**. I'm going to go through the activities within these stages, but I'll only lightly cover activities that you and I have lived through before, unless there are differences when applying the core chain method. Also, I will try to provide a more thorough introduction to activities that are relatively new (or are less substantial in the waterfall method).

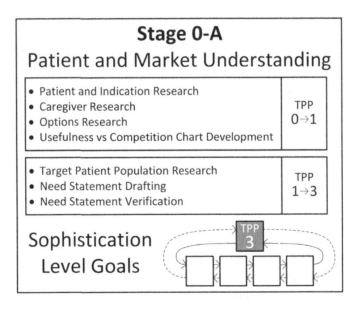

Stage 0-A: Patient and Market Understanding

THE CORE CHAIN

In this substage, we are going to do nothing other than live in the world of those we intend to help. We will be exclusively focusing on the TPP and need statement elements of the chain for this substage, as we want to build those up before we look too far down the chain. Following are some of the activities we can expect.

ACTIVITIES

Core Chain Sophistication Level Advancement: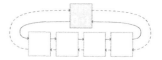

- **Patient and Indication Research**—To begin our journey, we will attempt to learn as much as we can about the patient and the indication through their eyes. This is not just an academic paper review; it is important to sit with patients and listen to their experiences and situations in their own words. We need to understand the issues they are experiencing, how it affects their lives, their support systems, their preferences for solutions, their concerns, and what led them to seek care.
- **Caregiver Research**—After we start to understand the patients' perspectives, it's time to turn our attention to the caregivers. The caregiver may be an individual or a team, and it may include the doctor, nursing staff, family members, physical therapists, etc. We will explore the indication and the patient from the eyes of these caregivers, and strive to understand the broader workflow that they fit into. We also need to understand the forces around the care, including finances, metrics they must meet, how they are personally incentivized, and how the broader system is incentivized. It's often like untangling Christmas lights, but absolutely worth our time.
- **Options Research**—We must understand all options available to the patients and caregivers, and in what situations those options change. We will also investigate the standard of care across the differences within the disease state and the current thinking about the selection of one option over another. Our discoveries will feed into our "line of best choices" for the usefulness vs. competition (UC) chart development.
- **Usefulness vs. Competition (UC) Chart Development**—The UC chart is our analysis of the landscape of choices available to the patient and caregivers. The chart is subjective, so we'll challenge our "usefulness" ratings with patients, caregivers, and administrators, and challenge our "competition" ratings with industry experts. We should be able to create the line of best choices and verify that line with caregivers and experts as the current standard of care.

Core Chain Sophistication Level Advancement: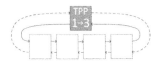

- **Target Patient Population Research**—We will now broaden our scope for the TPP to understand our core population, adjacent populations, and subpopulations within these population sets. We'll attempt to understand how these groups differ, and how their options and needs differ accordingly.
- **Need Statement Drafting**—We need to create a concise need statement based on our knowledge to this point. We'll then check that need statement against some knowledgeable parties.
- **Need Statement Verification**—We'll verify the need statement against the TPP and caregivers we intend to target. We won't just give them the statement and ask, "Does this look right?" We'll really challenge it, fostering objectivity as much as possible.

At the end of the patient and market understanding substage, we will be on firm footing with the left side of our core chain. We should have a strong understanding of the TPP as a whole, along with the patient and caregiver perspectives. We will also understand the array of options available to the patient and caregivers, and how useful those options truly are. We will have distilled a solid need statement, which has been verified for accuracy.

THE CORE CHAIN

Stage 0-B: Building the Case and the Plan

Moving into Stage 0-B, we'll have spent some time understanding the indication, our patients, the caregivers, and other technologies. So now, we'll dive right back into their world, validating our understanding and starting to convert that knowledge into user needs. We'll also start building the business case that we'll present to investors, gatekeepers, significant others, and so on, basically whomever can give you the money and resources to proceed.

ACTIVITIES

Core Chain Sophistication Level Advancement: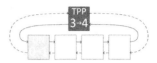

- **Target Patient Population Validation**—Any remaining characterization of the TPP should be done first, then we will proceed to validate our understanding of that TPP. This validation should be independent, ideally outside of our company (if we can swing it).
- **User Needs Efficacy and Basic Safety Drafting**—Our next big translation event starts here, as we begin to convert our understanding of the TPP, the caregivers, and their needs into the context of our field of expertise. We should end up with an itemized list that focuses on outcomes rather than solutions. We don't need to create all the user needs at this point, but a description of the basic outcomes we need to achieve and our major safety considerations should be documented.
- **Usefulness versus Competition (UC) Chart Verification**—We need to verify the beautiful UC chart we created in the previous substage. Once again, having an external party manage the verification is the cleanest approach, but at minimum, this can be verified by those determining usefulness (caregivers, administrators, and so on), and the line of best choices. The competition axis (see Chapter 19) can be challenged by a combination of industry analyses and an understanding of how "crowded" the different treatment options are to the caregivers.
- **Usefulness versus Competition Chart Analysis**—This is a fun activity, and it leads us into the business case development coming next. We will analyze the UC chart and determine where we can offer value to the patient, caregivers, and other stakeholders. This activity will visualize where our solution would potentially fit against other options, and what that means for strategy—for example, regicide approach versus fertile ground development (again, see Chapter 19).

Business Activities—No Core Chain Sophistication Level Advancement from These Activities

- **Business Case Development**—Ah, back to the good ol' business case. We have discussed this before, but there are some new considerations with the core chain method. First, we'll use the UC chart to show the area we feel we can target with a solution, and what usefulness we think we can achieve. We can relate that to the

THE CORE CHAIN

usefulness of the existing solutions, their prices, and how that can affect pricing strategies. If absolutely needed, we can create some basic concepts, but we must tread lightly so we don't bias them with a particular solution. We'll add bright flashing caveats that the solution and approach may (and will likely) change as the solution presents itself.

As always, there may be additional business activities we want to do in this stage (for example, building the team, picking a fancy logo for our business cards), and we can fill those in as we like. At the end of this stage, we will have a command of our patient populations, the indication, the systems they fit within, the options available to them, and the relative usefulness of these options. If our business case is persuasive, we have a good chance of moving forward. But as always, we must pass the gate first.

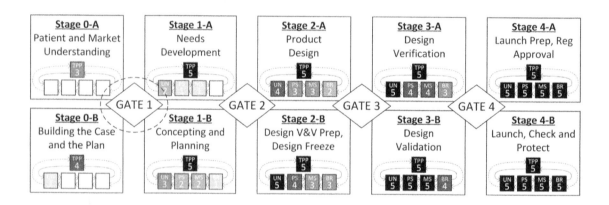

GATE 1

The gates are still present, but they have been tweaked a bit. I'm still standing in front of a group of gatekeepers as they poke and prod my data and dignity. But, as I mentioned before, they are going to evaluate things from a few different perspectives. We'll create a fancy presentation guiding them through these perspectives, first of the patient, then of the caregivers, then of the surrounding healthcare environment, and the resulting UC chart. We'll then walk through the products, usefulness and competition levels, as well as the opportunity that we see. We'll also show the economics of the project, overlay that with the economics of the business opportunity, and make the ask for resources. The gatekeepers will challenge us on our assumptions, including whether we're building the right chain or not. If we have executed the process well up to this point, we will be ready for the scrutiny. If everything is hunky-dory, they'll write a check and scoot us into Stage 1.

31

THE CORE CHAIN METHOD—STAGE 1

STAGE 1 OF the core chain method, aka the fun stage, is primarily about firming up the product requirements and starting to come up with concepts and plans to fulfill those requirements. Once again, we will have two substages that will break up our activities: **Needs Development (A)** and **Concepting and Planning (B)**.

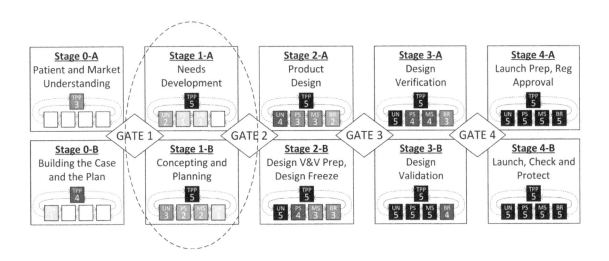

THE CORE CHAIN

```
┌─────────────────────────────────────────────┐
│              Stage 1-A                       │
│           Needs Development                  │
│ ┌─────────────────────────────────┬───────┐ │
│ │ • TPP and NS Review / Challenge │  TPP  │ │
│ │ • Need Statement Validation     │  4→5  │ │
│ │ • User Needs Development /      │       │ │
│ │   Initial Characterization      │  UN   │ │
│ │ • Intended Use Draft            │  1→2  │ │
│ ├─────────────────────────────────┼───────┤ │
│ │ • Concept Development Loop      │  PS   │ │
│ │                                 │  0→1  │ │
│ ├─────────────────────────────────┼───────┤ │
│ │ • Process Flow Creation /       │  MS   │ │
│ │   Process Planning              │  0→1  │ │
│ ├─────────────────────────────────┴───────┤ │
│ │ • Core Chain Trace Matrix Draft         │ │
│ │ • Concept Comparison                    │ │
│ │ • Build the Right Chain Review          │ │
│ └─────────────────────────────────────────┘ │
│                               TPP            │
│  Sophistication                5             │
│  Level Goals         UN   PS   MS            │
│                       2    1    1            │
└─────────────────────────────────────────────┘
```

Stage 1-A: Needs Development

In this substage we will continue our development of the user needs and start to make physical manifestations of the product. We are adding new elements to the core chain, the product specifications, and manufacturing specifications, so we must be conscious of translation errors. Here's an overview of what we need to do:

ACTIVITIES

Core Chain Sophistication Level Advancement: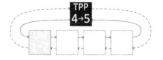

- **Target Patient Population and Need Statement (NS) Review/Challenge**—Before we dig too far into the user needs, we will take a minute and challenge our definition of the TPP and need statement. Both should be validated at this point, so we don't need

to spend much time on it; we just need to look for any biases or assumptions that could have led us astray.
- **Need Statement Validation**—Same as the TPP validation, this should be as independent as possible with a focus on removing bias or leading questions. In this validation, more patients and caregivers should be consulted than in the verification, and if a survey/questionnaire is used, it should be challenged to provide a relevant, unbiased view of the need statement accuracy.
- **User Needs Development/Initial Characterization**—In the last stage, we started listing out the user needs that we could easily think of, and now it's time to take a deeper look. We will work toward listing all of the safety and efficacy user needs, but we will likely have some questions. This activity will also inform our design inputs. As always, we will be careful to delineate, delineate, delineate.
- **Intended Use Draft**—Now that we better understand our patients and their needs, we can draft our intended use. This will be part of our final labeling and will likely set much of our regulatory pathway. We can leverage predicates here, and match language to similar devices if possible.

Core Chain Sophistication Level Advancement: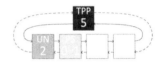

- **Concept Development Loop**—As we have done before, we will brainstorm potential solutions, exploring the entire design space. We will use our user needs and the UC chart as a road map of where to spend our effort exploring. We will create crude concepts and test them if we can.

Core Chain Sophistication Level Advancement: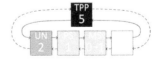

- **Process Flow Creation/Process Planning**—For each of our concepts (or in the instance that we have a pile of them, the concepts we're most likely to run with), we will build a theoretical process flow. This is a road map of the major process steps that may be needed, as far as we can create it. There will be some gaps in our process, as the product is nowhere near designed, but we need to start roughing it in, understanding if there are any major technical hurdles in a particular process.

Business Activities—No Core Chain Sophistication Level Advancement from These Activities

- **Core Chain Trace Matrix Draft**—In this substage, we can start to visualize the core chain we are building. How we do this is up to us, but we need to show how each of these elements trace back to one another, along with the verification, validation, and quality control efforts that prove out that trace. We will continuously update this core chain trace matrix throughout our development activities.
- **Concept Comparison**—Soon we will be amping up our development efforts, so we may need to take the many concepts and whittle them down to a handful in order to progress. We do this by evaluating the various concepts whichever way we can, including from the patient and caregiver point of view, as well as from the point of view of the business. We will consider the benefit and difficulty of each core chain that would result from each concept, their usefulness, difficulty, and risk throughout the chain. The concept comparison will be done in combination with the overall "build the right chain" review.
- **Build the Right Chain Review**—We will do this often, and rightfully so. The level of rigor will depend on us, but it is important that we challenge our path as we learn more. The review is a step-back review of the chain (at this point, *chains*—one for each concept), and how it relates to the patient, caregivers, and our business. We will spend most of our time building, so we must step back as we go and look at the overall picture, making sure it withstands our critical eye.

31—THE CORE CHAIN METHOD—STAGE 1

Stage 1-B
Concepting and Planning

• Full User Needs Review / Challenge • User Needs Verification	UN 2→3

• Design and Development Plan Creation
• Prototype Development / Testing Loop
  - Prototype Process Development
  - Prototype Manufacturing / Assembly
• Hazards and Harms Analysis
• Design Review
	PS
	1→2
	MS
	1→2
	BR
	0→1

• Build the Right Chain Review • Risk Management Plan • Initial IP submittal • Project and Business Plan Updates	

Sophistication Level Goals

TPP 5 — UN 3 → PS 2 → MS 2 → BR 1

Stage 1-B: Concepting and Planning

Coming out of the last substage, we should have some strong concepts that we are running with, hopefully not just one or two, but three, four, or more. We can advance them all, and let the cream rise to the top. But to keep them moving in the right direction, we'll start by advancing the user needs, and start building our design history file. After we have built and tested our prototypes, we'll get into the business side, updating the business plan, submitting some provisional patents, and performing our "build the right chain" review.

ACTIVITIES

Core Chain Sophistication Level Advancement: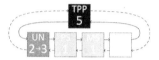

- **Full User Needs Review/Challenge**—We will go through the entire user needs list (and the other business needs and wants lists, if desired), and challenge it one last time before we start the process of verifying them.
- **User Needs Verification**—We will verify the user needs by somebody in the know, such as a subject matter expert, or some potential users of our device. We want an unbiased opinion before we put a ton of horsepower into achieving those needs.

Core Chain Sophistication Level Advancement: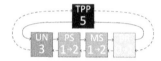

- **Design and Development Plan Creation**—As our user needs take shape and we go further into prototyping, it is time to create our official design and development plan. This document will lay out how we intend to develop, test, and control the product specifications.
- **Prototype Development/Testing Loop**—The second of three design iteration loops (concept, prototype, product), the prototype development/testing loop will progress our select concepts into more realistic prototypes. Our testing will advance in pace with the design, allowing us to challenge the different prototypes and refine our designs. Within this iteration loop, we will advance the manufacturing process and actual assembly process (batch record) by detailing all the process steps and equipment needed to produce each prototype. We will also assemble the prototypes using a basic assembly instruction, but likely with equipment that is only useful for prototype-level production.
- **Hazards and Harms Analysis**—Between this and the risk management plan, the hazards and harms analysis will kick off our official risk management activities. We're documenting the foreseeable potential harms associated with our device, along with the clinical hazards that could lead to those harms. It's still early with our device, but we'll start thinking about the risks associated with our solution.
- **Design Review**—We'll have a few of these, as they are part of the design control process. We will include the entire team here, representing all functional groups in the

development process, and create formal documentation of the meeting. The primary focus will be on our plans, the hazards and harms analysis, and the user needs.

Business Activities—No Core Chain Sophistication Level Advancement from These Activities

- **Build the Right Chain Review**—Once again, we'll take a step back and question whether we are building the right chain. We're going to do this a fair bit, so it is important that we do this well, genuinely and objectively challenging our approach based on our new understanding. We'll submit our assessment as part of the gate review, and the gatekeepers should tell us if we're full of beans.
- **Risk Management Plan**—Our risk management plan will describe how we will evaluate and categorize risk, and what documents we will deliver as part of the risk management process.
- **Initial IP Submittal**—We'll hopefully have a slew of provisional patents to submit at this point, starting the exclusivity clock and hopefully staying ahead of our competition.
- **Project Plan *and* Business Plan Update**—With our new knowledge of the user needs, the product, and the process, it is time to update the picture from both a project and business perspective. At this point we should have a clearer regulatory plan and a better understanding of the technical (and other) hurdles that lay in front of us.

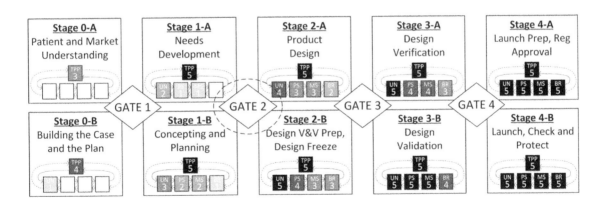

GATE 2

When we reach this gate, we'll be able to show the progression from our popsicle sticks into smooth-looking prototypes, and how the rest of the chain has developed. Gate 2 is probably the most fun review, as these gatekeepers get to actually see the wild concepts you came up with, and how those progressed into real prototypes. We can expect, however, that there will be a brutal "build the right chain" review, as they will get to see the broader design space, a better understanding of the user needs, and the more accurate economics of the business and project plans. This is good and we'll welcome it. If everything is aligned and the core chain is at the required sophistication levels, they'll let us pass into Stage 2.

32

THE CORE CHAIN METHOD—STAGE 2

STAGE 2 OF the core chain method has a strong focus on the product and the tests that challenge it. Compared to stages 0 and 1, this stage requires dramatically more work, takes longer, and is quite a bit harder. As we did in previous stages, we'll break this stage up into two major substages: **Product Design (A)** and **Design Verification and Validation Preparation/Design Freeze (B)**.

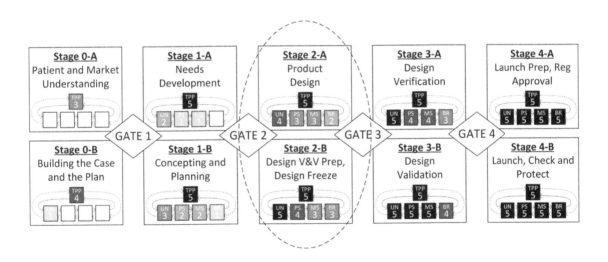

THE CORE CHAIN

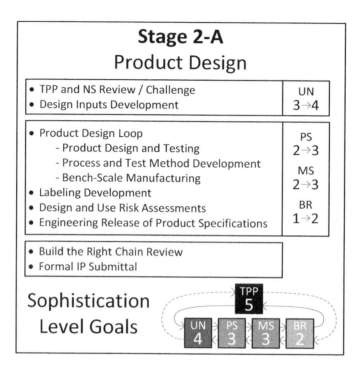

Stage 2-A: Product Design

The product design substage is one of the more significant substages, so we may be here for a minute. One item that you might note is that we are starting the main product development process without the user needs at a sophistication level of 5. That's intentional. We'll advance it to a 4 before we continue with our product design, but there is a high likelihood that we'll learn something about our user needs (and design inputs) during the product design process. We'll want to lock those user needs in *after* we get a bit smarter.

ACTIVITIES

Core Chain Sophistication Level Advancement: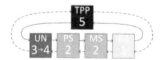

- **Target Patient Population and Need Statement Review/Challenge**—We'll do another kick of the tires on the TPP and need statement based on our new understanding of the user needs and the rest of the chain.

- **Design Inputs Development**—Done in concert with the user needs testing and characterization, we'll translate our high-level user needs into concrete, measurable design inputs. Let's keep in mind, though, that this is a translation event, and as such, the *prevent translation errors* core value applies. We'll also make sure to delineate, delineate, delineate.

Core Chain Sophistication Level Advancement: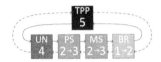

- **Product Design Loop**—This is our main product iteration loop nested within our overall core chain iteration loop. The product design loop is an iterative process that stretches across our product specifications, manufacturing specifications, and batch records. We will be repeating the loop of designing the product, updating the process, building the product, testing the product, and then starting over based on what we learn. We will sit in this loop until we sufficiently characterize the product and advance the core chain. We'll save the record of this activity as part of our design history file, and the output of this loop will be our "Device Master Record," comprising of our product specifications and manufacturing specifications.
- **Labeling Development**—At this point, we can develop drafts of the labeling for our packaging. The official labeling includes the instructions for use (IFU), artwork, labels, symbology, and basically anything else that is printed on or with the product.
- **Design and Use Risk Assessments**—During our product design loop, we will perform our official use and design risk assessments. Each of these will be leveled to what we know at that iteration, and we'll update them with detail as we know more. These two risk assessments will inform our design activities, so we will give them serious consideration.
- **Engineering Release of Product Specifications**—When we have a product that we feel is ready for more serious testing, we will snapshot that product through the product specifications. At this point, it's usually not a full release of the specs, but a release under engineering revision control. Some people like to call this the "design frost" as opposed to the "design freeze" that comes later.

THE CORE CHAIN

Business Activities—No Core Chain Sophistication Level Advancement from These Activities

- **Build the Right Chain Review**—That's right, we'll step back and look at the entire picture again, making sure there is a real and pressing need for what we're doing, and that we're addressing it properly.
- **Formal IP Submittal**—At this point we'll have a product to describe, so it'll be time to start converting our provisional patents into full patents.

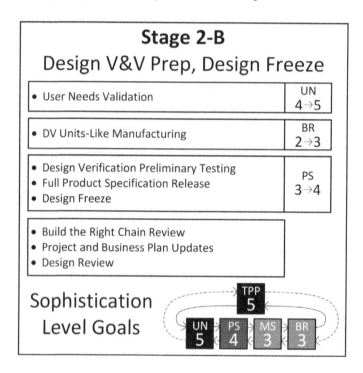

Stage 2-B: Design V&V Prep, Design Freeze

Coming into this substage we'll have a full set of engineering-released product specifications to work from, and we should know enough to finalize our user needs. We'll then test the product against what will be our verification tests, and at the same time we'll "test the tests" to ensure they are ready for primetime.

ACTIVITIES

Core Chain Sophistication Level Advancement: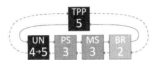

- **User Needs Validation**—The user needs validation will be a larger, independent confirmation of the user needs, sometimes performed by a third party. We'll create a protocol, test the user needs through subject matter experts, patients within our TPP, and caregivers.

Core Chain Sophistication Level Advancement: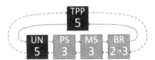

- **Design Verification (DV) Units-Like Manufacturing**—Using a scaled-down version of our final processes, we'll attempt to build product that meets our product specifications. We'll use an initial set of batch records and standard operating procedures to execute the process, challenging it as we build. There may still be some kinks in the process, but we're hoping to build units that will be similar to the ones we'll use for our official design verification later.

Core Chain Sophistication Level Advancement: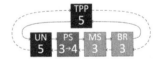

- **Design Verification Preliminary Testing**—This will be one last kick of the tires before we do our official design verification tests. We need to not only test that our product will perform as intended, but also that the tests that we are performing in the verification are mature enough as well.
- **Full Product Specification Release**—Once we have updated the product and we are confident that we are ready for full design verification, we will do a full product specification release. These are often the "Revision 1" documents, while the

THE CORE CHAIN

engineering release was "Revision A." Formal design control will kick in as soon as we release these specifications.

- **Design Freeze**—Once the full product specifications are released, the business usually calls a "design freeze" on the product, which is the professional equivalent of a teacher telling their class "pencils down." No more tweaking the product without invoking serious change control, so hopefully we got it right.

Business Activities—No Core Chain Sophistication Level Advancement from These Activities

- **Build the Right Chain Review**—Yep, again.
- **Project Plan *and* Business Plan Update**—Much of the picture should be coming into focus for us at this point, so we'll update the project plan and business plan with our latest and best information.
- **Design Review**—Through the product design loop and pre-verification testing, we will end up with a (hopefully) final set of product specifications that we can start verifying and validating. Before we do that, we'll bring everybody together and review what we came up with, challenging whether our product specs fulfill the design inputs and user needs. In preparation for this we'll update our core chain trace matrix, risk management documents, any project plans, and the design history file with our progress.

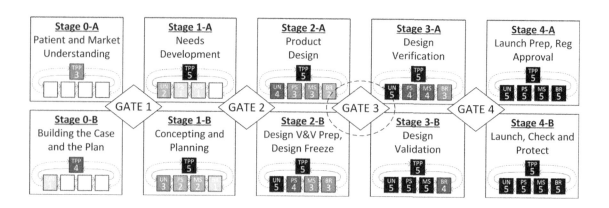

GATE 3

If we've done our job well, during this gate review we'll hand them a final(ish) product with some evidence showing that it does what we want it to do. They'll probably ask why it took us so long and cost so much, but, you know, they always do. With the product in hand and a manufacturing line in front of them, they'll again confirm that the project and business plans still make business sense.

33

THE CORE CHAIN METHOD–STAGE 3

In Stage 3 of the core chain method, we will have a design-locked product in hand, and we will perform the major design verification and validation activities. Those activities will be broken up into two substages: **Design Verification (A)** and **Design Validation (B)**.

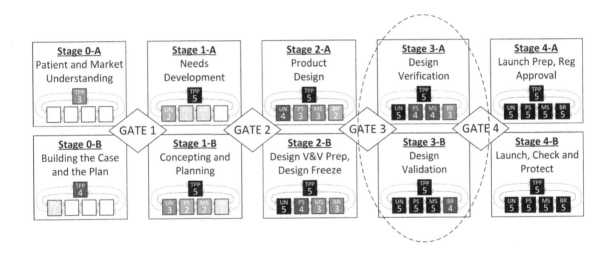

THE CORE CHAIN

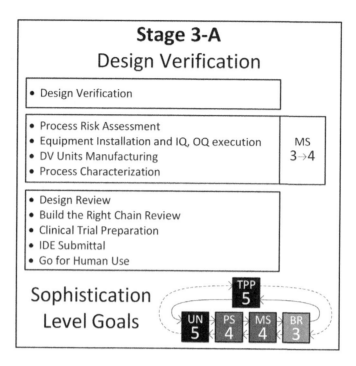

Stage 3-A: Design Verification

Our design verification efforts will dominate this substage, but they won't be all we do. We'll need to up the sophistication on our manufacturing and prepare for the clinical trial. The design verifications won't be done until the tests say they're done, so be ready for a sustained fast pace and coordination with a lot of external groups. Here's how it will unfold for us.

ACTIVITIES

Core Chain Sophistication Level Advancement: In Combination with the Design Validation Shown Later, These Activities Advance the Product Specifications

- **Design Verification**—We have covered what we need to do here, so I won't bog you down with the details. (I can tell your eyes are starting to glaze over just thinking about it.) One reminder, though: we are testing the product specifications against the design inputs here, not the actual user needs. We will perform exhaustive bench testing of our product during the verification, including performance, durability, usability, biocompatibility, and packaging validation activities.

33—THE CORE CHAIN METHOD—STAGE 3

Core Chain Sophistication Level Advancement:

- **Process Risk Assessment**—We'll perform our full process-based risk analysis, which assumes the product was designed correctly, was used correctly, but was built incorrectly. We'll see how failures in our process translate to harms to the patient, and ensure that we have the proper controls in place.
- **Equipment Installation and IQ, OQ Execution**—At this point (or earlier in some cases), we should be installing and validating our final production equipment, albeit not necessarily at final capacity. We will ensure the equipment is installed and is operating correctly before we do our main process characterization.
- **DV Units Manufacturing**—These units will be the most official ones we've made to date. Many of our quality system processes will kick in for this build. We'll need lot control, operator training, batch records, and so on. These units will be used for our design verification and some (nonclinical) design validation testing.
- **Process Characterization**—We will be testing the range of the critical variables for each process, understanding how they affect the product's compliance to the product specifications. Keep in mind "the smaller the links, the tougher the chain" concept here, as we want to document the ties between different process variables and individual elements of the product specifications.

Business Activities—No Core Chain Sophistication Level Advancement from These Activities

- **Design Review**—This is our last design review before human use, so we will be thoughtful and careful before moving on. At this point there will be pressures galore to get this project moving, to meet deadlines, to not lose another day or another dollar on the project. We'll take a breath, look at the data, and ask ourselves if we are okay with this product going into our own bodies or into someone we care about. If we can answer that question with a *yes*, and we pass our design verification, we'll move forward. If not, we'll keep at it until it's right.
- **Build the Right Chain Review**—Hopefully, this is a minor activity at this point. But we'll again be smarter than the last time we did the review, and the competitive

landscape may have changed. So, we'll step back and review the chain in the context of the landscape once again.

- **Clinical Trial Preparation**—Institutional review boards (IRBs), clinical site management, clinical protocols, contract research organizations, oh my. This is as much fun as it sounds, and we'll build a subset of the team to do nothing but this work.
- **Investigational Device Exemption (IDE) Submittal**—Depending on the product, the location of the IDE submission varies, or it's not required at all. For our product, we will submit it here. If accepted (or more precisely, not rejected), we can start our human trials.
- **Go for Human Use**—This is one last check by the people for whom the buck stops.

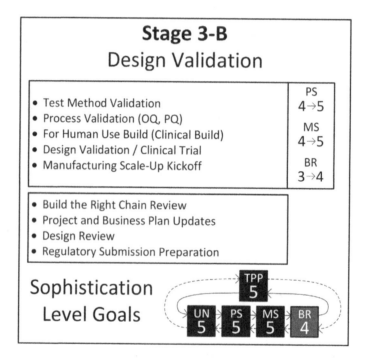

Stage 3-B: Design Validation

If we need a clinical trial, and with our product we likely will, the pivotal trial occurs here. We will be sending our product out into the wild and seeing how it does. The trial(s) will likely take a while, so in the meantime we will continue our preparations.

ACTIVITIES

Core Chain Sophistication Level Advancement: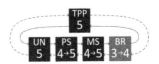

- **Test Method Validation**—We need to test the tests that we intend to use in our final manufacturing. Test methods are crucial and generally underappreciated, so we'll make sure they're up to snuff before counting on them to tell us whether our product can leave the facility.
- **Process Validation (OQ, PQ)**—Once we have that understanding of how the process variables affect the product, and their allowable ranges and sweet spots, we'll build product at the extremes of the ranges (OQ) and the sweet spot (PQ). We will do it multiple times to show lot-to-lot variability, and we will show that as long as we stay within the dictated process ranges, we will make product that meets the product specifications.
- **For Human Use Build (Clinical Build)**—This will be a serious milestone for our operations, and it's not something they'll take lightly. We'll implement good manufacturing practices, and be as close to our commercial manufacturing (although possibly not at the same scale) as we can be at this point. There will be a formal review of this product and special labeling before we'll release it for the clinical trial.
- **Design Validation/Clinical Trial**—If we've done our job well, we shouldn't learn much during the clinical trial. The clinical trial is our most blunt tool to test our core chain. If we've built and validated the individual elements correctly, we should learn nothing new by testing them all at once. The clinical trial(s) is the last element of our design validation.
- **Manufacturing Scale-Up Kickoff**—At this point, our equipment is installed and our processes are validated, but we need to be ready to meet demand at launch and after. We'll start our scale-up efforts in this stage as we wait for the clinical trial to complete. And we'll continue our efforts as we wait for regulatory approval in the next stage. Scaled-up equipment takes a long time to build after all.

Business Activities—No Core Chain Sophistication Level Advancement from These Activities

- **Build the Right Chain Review; Project Plan, and Business Plan Updates**—The usual suspects.

THE CORE CHAIN

- **Regulatory Submission Preparation**—As we prepare for the clinical trial to wrap up, we will prepare our regulatory submission documentation. This may be for a 510(k), PMA, or a *de novo*, as they each have their own set of submission requirements. All medical device submissions align with the core chain logic, so this should be more a matter of putting pieces together rather than the creation of new items for the FDA.
- **Design Review**—This is our final design review, and it will signal the culmination of years of hard work. We will focus on the results of the clinical trial, if the core chain is fully complete and traced, and whether we think the product is safe and effective enough to be released into the hands of the clinicians.

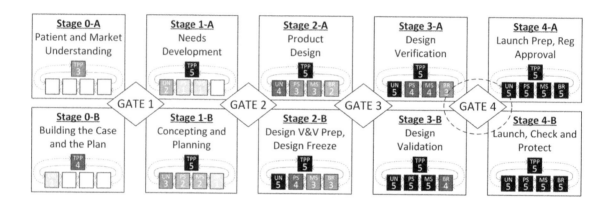

GATE 4

If we make it to this gate, we are in the rarefied air of medical device development projects. Asking to move on to Stage 4 means our product has made it through design verification and validation and has passed the clinical trial, and we have a production line raring to go. The gatekeepers will again grumble about money and time, but I'm sure we'll find a twinkle in each set of eyes. Things will be hopeful if we get to this point, but there will be one last hurdle to overcome, and it's a doozy.

34

THE CORE CHAIN METHOD–STAGE 4

STAGE 4 OF the core chain method will mark the end of our product development journey, for this product at least. We'll still need to make it through regulatory approval, and our official launch, but after that, we'll switch to "check and protect" activities. Here are the substages we will see in this stage: **Launch Preparation, Regulatory Approval** and **Launch, Check and Protect**.

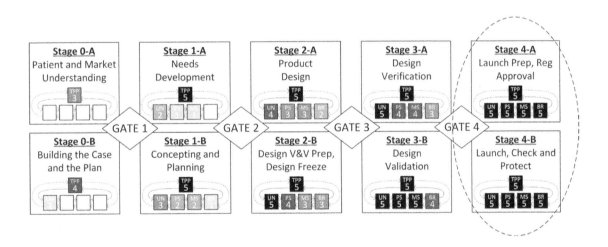

THE CORE CHAIN

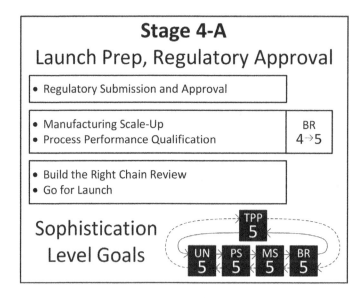

Stage 4-A: Launch Prep, Regulatory Approval

We will submit our regulatory submission to the FDA and start the review clock. They'll let us know they got it and will then go dark for a while. In the meantime, we'll get ready for launch on the assumption that the approval is coming. How ready we get will be based on our confidence that we'll get approval (and confidence in the approval timeline).

ACTIVITIES

No Core Chain Sophistication Level Advancement

- **Regulatory Submission and Approval**—We will write up our core chain, along with its development history, in the format that the FDA expects, put a bow on it, and send it in. Like a parent sending a child to kindergarten on the first day of school, we'll have tears in our eyes and wish we could go too, protecting them and telling everyone about our little treasure. We'll then wait by the bus stop for at least three months (510[k]), or six months (PMA), until our little being reemerges, with more questions than we'd like.

Core Chain Sophistication Level Advancement: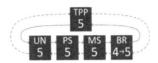

- **Manufacturing Scale-Up**—As our product nears approval, we get to decide how much we want to count on its regulatory clearance, and on our sales projections. Even if we're confident on the clearance, as far as our sales projections go, those tend to be . . . let's say a little optimistic. We'll likely land on cautiously optimistic, with a plan to speed up our scale-up if certain sale-goal triggers are met. This activity includes additional equipment, operators, supplier preparation, inventory build-ups, and so on.
- **Process Performance Qualification**—Once our production is in its launch state, we will execute our process performance qualification, which tests our entire manufacturing and distribution processes. We will run multiple lots, send the product around the nation and back to us, then test the heck out of it. If it looks good, we'll say our manufacturing is in a validated state, and ready for the big time.

Business Activities—No Core Chain Sophistication Level Advancement from These Activities

- **Build the Right Chain Review _and_ Go for Launch**—Once we get our regulatory approval or clearance, we are allowed to release it to the marketplace. While we have been given the go-ahead by the powers that be, we will take one last "go for launch" check around the table, NASA-style.

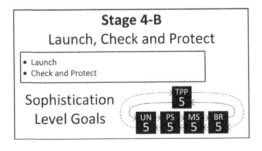

Stage 4-B: Launch, Check and Protect

THE CORE CHAIN

This substage is what we have all worked toward for years, and hopefully, it means that the money starts traveling inward rather than just outward. More important than that, though, it means that the product that we so tirelessly developed will start making it into the hands of the clinicians and patients that need it. If we did our job well, the need statement will start to be fulfilled. The standard of care will improve, the options to patients and caregivers will increase, and we will directly help our TPP, the all-important target patient population. Arrival into this substage marks the end of our product development process. There will be streamers, champagne, and war stories. Then we'll take a breath and get back to work.

I know when we lay out the entire journey ahead of us it can feel daunting, but you and I both know how exhilarating it can be. This is especially true when you consider the real impact we may be able to make on the lives of real people. With the core chain method as our approach, I feel that we have the best shot at getting this product in the hands of caregivers. Are you ready? Great, then let's get started.

35

PART IV: RECAP

There's lots I want you to remember from Part IV, but here are some of the highlights:

KEY CONCEPTS

The core chain method is built around *visualization*, *alignment*, and *iteration*.

- Visualization—We visualize the progression of the core chain through our *sophistication levels* and level up individual elements of the core chain through various product development activies.
- Alignment—We *build left to right*, which means we always need the same or higher sophistication on the core chain element to the left.
- Iteration—We *rough it in and refine*, which means we increase sophistication from the left, but then continue down the chain, rolling out our new understanding, again and again, with increasing levels of sophistication.

THE CORE CHAIN

Overview of the Core Chain Method of Medtech Development

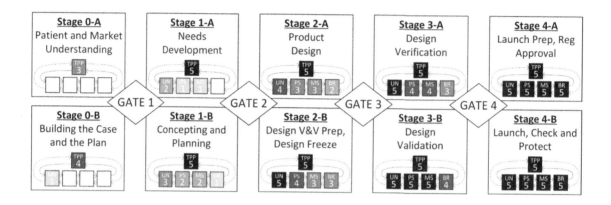

Reference Set of Sophistication Levels of the Core Chain Elements
(Remember, adapt these to your project)

	Target Patient Population (TPP) Sophistication Levels
1	General target patient population identified and described.
2	Typical patient profile created. First draft of need statement created.
3	Target patient population well defined with clear borders. Possible subpopulations and patient journey identified. Need statement challenged and further crafted.
4	Target patient population characterized with defined subpopulation segments with profiles. Target patient population definition is validated.
5	Final target patient population is thoroughly characterized, described, and validated. Need statement is finalized and validated.
	User Needs (UN) Sophistication Levels
1	Primary efficacy and indication-specific safety user needs defined but potentially non-specific. Non–core chain requirements delineated off user needs list.
2	More complete efficacy and safety user needs list is created. Initial testing/characterization of user needs.
3	Regulatory and statutory requirements added. Efficacy and safety needs more deeply characterized with light verification by SMEs or individuals within the TPP. Initial design inputs are drafted.

4	All user needs are fully defined and specific. Borders of individual user needs challenged. user needs have been verified against the TPP and associated caregivers. Design inputs are fully detailed.
5	All user needs are fully defined and have been objectively validated against the TPP and associated caregivers. Design inputs are verified against the user needs.

Product Specifications (PS) Sophistication Levels	
1	Popsicle sticks and pipe cleaners. Looks like you made it in kindergarten. Possibly some digital concepts.
2	More sophisticated concepts. Basic prototyping methods (3-D printing, and so on) to create physical representations of the product. The specific concepts are tested for specific types of product feedback.
3	Physical representations (although using some different processes than the final production processes) have been created. Product specifications are complete. Product has not been fully characterized.
4	Product has been built and tested using scaled-down versions of the final production processes. Product specifications are fully detailed but not fully challenged.
5	Final product specifications, including software designs, are complete, verified, validated, and frozen. Actual production-equivalent units of the product exist.

Manufacturing Specifications (MS) Sophistication Levels	
1	Basic process flow created showing all the major steps involved in making the product.
2	All process steps captured and possible equipment and process materials defined. Major process risks identified.
3	Detailed process flow created with all materials ins, outs, process variables identified. Test methods identified. Preliminary equipment defined. Pilot line created.
4	All processes and final equipment defined. Final equipment installed and IQ'd/OQ'd. All processes and equipment fully characterized. Test methods fully defined and described.
5	All individual processes, equipment, and test methods fully developed, characterized, and validated.

Batch Record (BR) Sophistication Levels	
1	Basic process flow walked through by operations/experts.
2	Basic assembly captured and attempted—not necessarily on production equipment.
3	Initial batch record/work instruction set created and tested by operators on early production equipment or surrogates.

THE CORE CHAIN

| 4 | Full production batch records/work instructions created, executed, and challenged by operators and engineers on production equipment. |
| 5 | Full production batch records/work instructions created, executed, and challenged by operators and engineers on production equipment. Entire production PPQ'd. |

CONCLUSION

Better we raise our skill than lower the climb.—Royal Robbins

On a Saturday in early September 2020, Jaymie Shearer and three friends met at the Mammoth trailhead in central California. They were there to celebrate the birthday of one in the group by embarking on an eight-day hiking and climbing expedition through the Sierra National Forest.

California wildfires were being battled in central California, but the hikers were aware of them, and had planned their route to steer well clear of the more than two dozen fires raging across the state. However, unbeknownst to them, a new fire in the area had started overnight, a particularly fast-moving and destructive blaze that would eventually be known as the Creek Fire.

Shearer and her friends set out into the Sierra Forest packed with enough supplies for their eight-day journey through the famed Ansel Adams Wilderness. As the day drew on, the light smoke they had expected seemed to be growing ever thicker. Continuing deeper into the forest, they began to notice that the color of the sky seemed to be changing, first to orange, then to grey. It became harder to breathe, and they decided to stop at an upcoming bluff and assess the situation. As they looked out, they could see a pyrocumulonimbus (fire cloud) forming over the forest, and they could hear its rolling thunder. Using their satellite phones, they began to contact people to try to get a read on the situation. They were only able to get scant details, but they learned about the new blaze in the area, and that they were probably headed right into it.

They decided to turn back and try to make it to one of their cars parked at a different trailhead. As they made their way through the forest, the situation appeared to grow more dire, and they became more and more nervous. The friends started discussing other options, including racing back. Shearer, an experienced hiker and climber, calmed her friends by repeating a famous hiking saying: slow is smooth, and smooth is fast. The friends adopted the mantra, continuing at a steady pace down the path.

They eventually made it to their car but learned that the main bridge leading them out of the area had collapsed. They began to carefully set out on the road but found that there were cars speeding past them in both directions, honking and flashing their lights. They tried to talk to people, but everyone had a different narrative about what was happening. Confusion seemed to be surrounding them.

They decided to make a clear plan and methodically stick with it. Slow is smooth, and smooth is fast. Getting back out of the car and taking supplies for three more days, they set out toward Devils Postpile Monument. At the monument was another of the group's cars, as well as access to a different road out of the area. Getting to the monument, however, required them to cross 13 more miles of forest. They proceeded back into the forest, toward the distant monument. Four hours in, they camped for the night, wondering if they should press the SOS button on their satellite phone, but decided that since they were moving at a steady pace, they could wait. They continued on, repeating their mantra, and periodically stopping to rest and rehydrate. Almost 24 hours after they had started the last leg of their journey, through a thick fog of soot, they emerged out of the forest and into Red's Meadow, finding their car. They grabbed the premixed mai tai cocktails in the trunk, then drove out of the area to safety. The Creek Fire continued to rage for another three and a half months, eventually consuming over 375,000 acres, making it the fifth-largest wildfire in modern California history.

Shearer and her friends were in a scary, complex, and rapidly changing situation. A natural response is to travel at top speed in any direction that seems to lead away from the danger. In fact, this is what most people around them were doing. Despite this, Shearer's group was able to slow down, understand where they needed to go, then use their principles to keep them on track. This may have saved their lives.

While medical device development is not a raging forest fire, it is a wilderness. The stakes are still high, and it's notoriously easy to wander off track. Even if we're excited to get started, it is important to slow down, understand exactly where we are going, how we will get there, and how we intend to stay on track.

With that in mind, I would like to stop and take a moment to thank you for reading this book. There is a lot of information in it, and I appreciate you taking the time to delve into its many concepts with me. There is much to consider, as the world of medical device development is a complex one.

At the very beginning of the book, you may have noticed the subtitle "A Survival Guide to the Perilous World of Medical Device Development." This is what the book is intended to be, a guide to medtech development along with the tools you'll need to navigate it. As we are coming to the end of our journey together, let's take a moment to recap where we've been.

In Part I, we went through the typical process of developing a medical device, then examined how and why it is often unsuccessful. In Part II, I introduced you to the *core chain*,

CONCLUSION

which is the ultimate goal of our journey, and where we must aim to lead ourselves through the wilderness. In Part III, we discussed guiding principles to keep you on path during your journey, which I call the *core values*. In Part IV, we created the path itself, to lead us from the very beginning to the very end of our journey, step by step. These tools are now at your disposal, and I hope the path in front of you is clearer, and more likely to be successful.

Well, mountaineers, off we go, into the Alpines. As we pause to reflect on the wilderness ahead of us, I hope we're doing so with fresh eyes. We know that medical device development is daunting, but the reward is worth the work, especially if we're smart and efficient in our approach. If we use the core chain as our goal, the core chain method as our path, and the core values as our principles, the wilderness should seem imposing yet conquerable. And while any medical device development that improves care is worth doing, I hope we can use our newfound approach to go after the big peaks, the transformative technologies. After all, the best view comes after the biggest climb.

As you set off, please keep in mind that you are not embarking alone, but as part of a community. There is a quote from Roman Payne in *Rooftop Soliloquy*: "I wandered everywhere, through cities and countries wide. And everywhere I went, the world was on my side." If you are in medical technology development, I am on your side, and those who I count as my close colleagues would say the same. We in the medtech community are here to help, as your success helps us all.

Again, I'd like to thank you for reading this book and for your involvement in our industry. The patient is the point, after all, and we're lucky enough to be in a position to help.

One last thing, a bit of advice from mountaineer Dan May: "When preparing to climb a mountain—pack a light heart."

FREQUENTLY ASKED QUESTIONS

As I'm sure you've noticed, delving into the intricacies of medical device development can get a bit tedious, but the stakes are quite high. I appreciate you sticking with it. I'm guessing you had plenty of thoughts and questions along the way, but I'm hopeful that the concepts can help guide you to an answer, whether alone or with your team.

Now, if you found yourself saying, "Whoa, bub, not that simple," well then, I'm guessing you're an engineer. I am also, by training at least, an engineer. As with every engineer I have ever met, there is a deep-seated compulsion to provide the complexity of the entire picture at the same time you are explaining the basic structure. This is a wholly peculiar trait and is a surer way to peg someone as an engineer than by looking at their business card. This often manifests itself as a lecture on theoretical physics when someone, with a very mild and fleeting interest, asks how barcode scanners work. Or, in another common manifestation, not ever saying, or allowing anybody else to say, anything definitive.

In the interest of clarity, readability, and good taste, I did my best to battle that compulsion. The result is that sometimes I had to sacrifice complexity for clarity. I hope you will be able to layer on the complexity, in both theory and practice, as you build on the foundation that this book provides.

There are some questions, however, that frequently come up as I discuss the core chain methodology, which I will go through here. If you have any additional questions beyond what I discuss below, feel free to contact me at justin@thecorechain.com.

Can the core chain show the locations of the Design History File (DHF), Device Master Record (DMR), and Device History Record (DHR)?

Sure can—here's a handy illustration showing the DHF, DMR, and DHR positions in the core chain.

THE CORE CHAIN

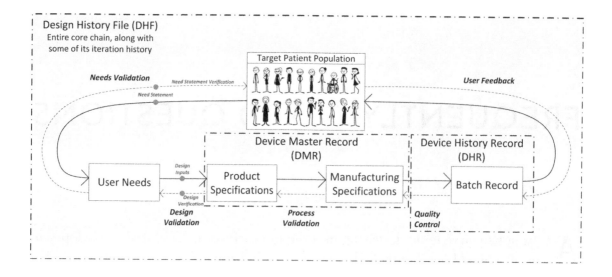

How is this different for big companies compared to start-ups or smaller companies?

The quick answer is the core chain is the same, but the lens through which someone considers it is often different. All the core values are important in both situations, but people in larger companies will find some of the core values harder to implement than those in a start-up, and vice versa.

In a big company, roles are typically well defined, work is more siloed, and inventor's bias is less of a concern. In the place of inventor's bias, however, is the "law of the instrument" bias, that is, "To a person with a hammer, all problems look like nails." When a company has an entire portfolio of stents, it's hard not to consider every hollow organ problem as solvable by stents. This is not inherently bad, as we can discover new uses for existing technologies, but the risk is that clinical problems are not looked at objectively. One of my consultant friends talks about this sometimes leading to a "push versus pull" mentality. Rather than hearing what the patient and caregivers truly need, it is pushing a solution that works best for your company. That can create a culture of iterative thinking, lower risk, and lower-benefit launches. To battle this, the core values "build the right chain" and "build from left to right" should be the focus.

The other common issue with larger companies is the specialization of their people, and the silos that often result. This matches the sequential waterfall style of development, where the project is handed off to different groups that build the core chain entirely from left to right, and then complete each section before moving on. While these companies get very good at these handoffs, much is lost with this approach. Focusing in on the "rough it in and refine" approach, with a cross-functional team that is cocreating, is important to improve their medical device development process. Luckily, these companies typically have the resources to execute that approach from the beginning, turning people's specialization into a pro rather than a con. The problem is, the company culture is typically more embedded, and a larger, more concerted effort must be made to change it.

In a start-up, you don't have to worry much about embedded culture. You get to instill a culture onto a blank canvas. You have an entirely different set of issues, however, and a different focus on the core chain as a result. We have belabored "inventor's bias" as an issue, and with good reason. As a large company must focus on "build the right chain" and "build from left to right" while killing the "law of the instrument" bias, so too must a start-up focus on the same core values, but while eliminating their inventor's bias. This is tough in practice but is fundamental to improving the success rate of these companies.

Another concern for smaller companies is how much effort they can put into the thoroughness of the core chain they are building. The core value, "the smaller the links, the tougher the chain," requires a deep characterization of sections of the chain against the previous section, and a line-by-line tie-out between the two if possible. The "prevent translation errors" core value also requires a deeper analysis of each section of the chain, all while money is tight and the clock is ticking. While it is often hard for a start-up to focus on these efforts, it is the *marrow* of what you're building. If they are glossed over, everything that is the essence of your company becomes weaker.

As the culture is more embedded, a larger effort must be made to change it.

What about software in the medical device (or as the medical device)?

As you may know, when we talk about software in the context of medical devices, there are two types: Software *in* a Medical Device (SiMD) and Software *as* a Medical Device (SaMD). SiMD is software as part of a software/hardware solution (think pacemaker). SaMD is a medical device that is nothing but software (think software that reviews images to help detect breast cancer). We approach them a little differently from a development perspective, but the principles of the core chain and the development process remain unchanged. In the case of the SiMD, we still have the same elements of the core chain, we'll just add specific activities and documents for software. For instance, in addition to the design inputs, we'll add software requirements, and in addition to our other risk analyses, we'll add a software-specific one. In the case of SaMD (pure software), you replace some of the activities with software-specific ones, that is, software requirements *instead* of design inputs. The level of documentation depends on the risk, but it should still fit the core chain framework nicely.

THE CORE CHAIN

This book is focused on medical devices. Does the core chain work for pharmaceuticals or biotechnology products?

A lot of the same principles apply, but the overall approach is a bit different. Medical device development is a logical, physical series of cause and effect. You have a dysfunction, you define it, you define what a solution would accomplish, you build that solution, then you prove that the solution accomplished what you said it would. Effects on the body are usually quite measurable and predictable. Pharmaceutical and biotechnology development is different. I believe they would like to be able to use the same approach, but what is going on at the fundamental level within the body is more difficult to understand and predict. They rely on *empiricism* more. Quite early in their development process, they start putting the product into animals and people, then they see what happens. With a medical device, that step happens much later in the process, because it can happen much later. In pharma, you work backward from empirical data to create a theory, and you back it up by the sheer number of subjects tested. With a medical device, you create a logical theory, then prove its validity with empirical data. Put simply, medtech is logic backed by data, and pharma is data backed by logic.

As a result, medical device development has more freedom. If you know the theory is sound, you have more freedom to change things so long as the change follows the same logic and is still backed by the data. In pharma and biotech, things follow a more "captured magic" approach, where all data is invalid if you change anything, since we still don't really understand what's happening at the fundamental level.

But even with those differences, the logic of the core chain can still apply. Pharmaceuticals and biologics use the quality by design (QbD) approach to development, which has similar elements to the core chain. There is a nice traceability between the critical quality attributes, the critical process parameters, and the corresponding control plan. It's also notable that the "rough it in and refine" core value is hardly needed, as it is redundant to the standard approach. The product and the process are more tightly linked in pharmaceuticals and biologics, a common phrase being, "The product is the process" during development. You can't try the product without advancement of the process to the same level. The other core values generally apply, but again, the lens is a bit different.

GLOSSARY OF ACRONYMS AND TERMS

510(k)—Premarket submission made to the FDA to demonstrate that the device to be marketed is as safe and effective, that is, substantially equivalent, to a legally marketed device.

BR—batch record.

CDRH—Center for Devices and Radiological Health (US FDA).

CRO—contract research institution.

dFMECA—design risk assessment/design failure mode, effects, and criticality analysis.

DHF—design history file.

DHR—device history record.

DMR—device master record.

DV—design verification.

FDA—Food and Drug Administration of the United States.

FIH—first-in-human.

FMEA—failure mode and effects analysis.

HIPAA—Health Insurance Portability and Accountability Act, generally known as a privacy rule that establishes national standards to protect individuals' medical records and other individually identifiable health information.

IDE—investigational device exemption.

IFU—instructions for use.

IP—intellectual property.

IQ—installation qualification.

IRB—institutional review board.

THE CORE CHAIN

Medtech—medical technology. I often use it as a surrogate for medical devices, but it is a bit broader than that and can encompass software and other systems that impact patient care.

MS—manufacturing specifications.

NS—need statement.

OQ—operational qualification.

pFMECA—process risk assessment/process failure mode, effects, and criticality analysis.

Pharma—pharmaceutical development industry.

PMA—premarket approval application.

PPQ—process performance qualification.

PQ—performance qualification.

PS—product specifications.

SaMD—software as a medical device.

SiMD—software in a medical device.

SME—subject matter expert.

SOP—standard operating procedure.

The Strategics—the largest medical device companies.

SWOT—strengths weaknesses opportunities threats (analysis).

TLA—three letter acronyms.

TPP—target patient population.

UC Chart—usefulness versus competition chart.

UN—user needs.

V&V—verification and validation (aka design verification & validation).

Made in the USA
Middletown, DE
01 September 2024